Macromolecular Drug Delivery

METHODS IN MOLECULAR BIOLOGY™

John M. Walker, SERIES EDITOR

METHODS IN MOLECULAR BIOLOGY™

Macromolecular Drug Delivery

Methods and Protocols

Edited by

Mattias Belting

Lund University, Lund, Sweden

 Humana Press

Editor
Mattias Belting
Lund University
Lund, Sweden
mattias.belting@med.lu.se
belting@scripps.edu

Series Editor
John M. Walker
University of Hertfordshire
Hatfield, Hertfordshire
UK

ISSN: 1064-3745   e-ISSN: 1940-6029
ISBN: 978-1-58829-999-4   e-ISBN: 978-1-59745-429-2
DOI 10.1007/978-1-59745-429-2

Library of Congress Control Number: 2008937921

Cover illustration: From Chapter 12, Figure 3D

Printed on acid-free paper

springer.com

Preface

Macromolecular drugs hold great promise as novel therapeutics of several major disorders, e.g., cancer and cardiovascular disease. However, their use is limited by lack of efficient, safe, and specific delivery strategies. Successful development of such strategies requires interdisciplinary collaborations, which provide opportunities for the unexpected at the interface between different disciplines. Once available, macromolecular drugs will revolutionize the treatment of various diseases as well as provide novel diagnostic tools for the benefit of the patient.

This volume gives an introduction to macromolecular drug delivery and a comprehensive review of the methods used in the field. In vitro and in vivo models are described that should be of interest to a broad scientific audience, including experts in biophysics, chemistry, cell biology, and pre-clinical and clinical in vivo studies. Researchers involved in the rapidly expanding field of targeted therapy strategies, e.g., in oncology, should also find this volume highly interesting.

Contents

Contributors

RACHIDA ABES • *UMR 5235 CNRS, Université Montpellier 2, Place Eugene Bataillon, 34095 Montpellier cedex 5, France*

SAÏD ABES • *UMR 5235 CNRS, Université Montpellier 2, Place Eugene Bataillon, 34095 Montpellier cedex 5, France*

DANIEL G. ANDERSON • *Department of Biological Engineering, Department of Chemical Engineering, and Center for Cancer Research, Massachusetts Institute of Technology, Cambridge, MA 02139, USA*

YULIA ANTOV • *Department of Physiology and Pharmacology, Sackler Faculty of Medicine, Tel-Aviv University, Ramat Aviv, 69978 Tel-Aviv, Israel*

ANDREY A. ARZUMANOV • *Medical Research Council, Laboratory of Molecular Biology, Hills Road, Cambridge CB2 2QH, UK*

ALEXANDER BARBUL • *Department of Physiology and Pharmacology, Sackler Faculty of Medicine, Tel-Aviv University, Ramat Aviv, 69978 Tel-Aviv, Israel*

MATTIAS BELTING • *Department of Clinical Sciences, Section of Oncology, Lund University, Sweden*

CARL A.K. BORREBAECK • *Department of Immunotechnology, Lund University, CREATE Health – Center for Translational Cancer Research, Lund, Sweden*

MICHAEL BUR • *Saarland University, Biopharmaceutics and Pharmaceutical Technology, Saarbruecken, Germany*

NICOLE DAUM • *Saarland University, Biopharmaceutics and Pharmaceutical Technology, Saarbruecken, Germany*

ANTONIA DIMITRAKOPOULOU-STRAUSS • *Medical PET Group, Clinical Cooperation Unit Nuclear Medicine, German Cancer Research Center, Heidelberg, Germany*

JOHAN FRANSSON • *Centocor Discovery Research – San Diego, 3210 Merryfield Row, San Diego, CA 92121, USA*

MICHAEL J. GAIT • *Medical Research Council, Laboratory of Molecular Biology, Hills Road, Cambridge CB2 2QH, UK*

PRAFULLA C. GOKHALE • *Departments of Radiation Medicine and Biochemistry, Molecular and Cellular Biology, Georgetown University Medical Center, Washington, DC 20057, USA*

BAGAVATHI GOPALAKRISHNAN • *Mirus Bio Corporation, 505 S. Rosa Road, Madison, WI 53719, USA*

MARSHALL GRANT • *MannKind Corporation, 1 Casper Street, Danbury, CT 06810, USA*

JORDAN J. GREEN • *Department of Biological Engineering, Department of Chemical Engineering, and Center for Cancer Research, Massachusetts Institute of Technology, Cambridge, MA 02139, USA*

ASTRID GRÄSLUND • *Center for Biomembrane Research, Department of Biochemistry and Biophysics, Stockholm University, SE-106 91 Stockholm, Sweden*

KULLERVO HYNYNEN • *Department of Medical Biophysics, University of Toronto and Sunnybrook Health Sciences Centre, 2075 Bayview Ave, Toronto, ON, Canada*

GABRIELA D. IVANOVA • *Medical Research Council, Laboratory of Molecular Biology, Hills Road, Cambridge CB2 2QH, UK*

USHA N. KASID • *Departments of Radiation Medicine and Biochemistry, Molecular and Cellular Biology, Georgetown University Medical Center, Washington, DC 20057, USA*

RAFI KORENSTEIN • *Department of Physiology and Pharmacology, Sackler Faculty of Medicine, Tel-Aviv University, Ramat Aviv, 69978 Tel-Aviv, Israel*

KELLY S. KRAFT • *MannKind Corporation, 1 Casper Street, Danbury, CT 06810, USA*

ROBERT LANGER • *Department of Biological Engineering, Department of Chemical Engineering, and Center for Cancer Research, Massachusetts Institute of Technology, Cambridge, MA 02139, USA*

BERNARD LEBLEU • *UMR 5235 CNRS, Université Montpellier 2, Place Eugene Bataillon, 34095 Montpellier cedex 5, France*

CLAUS-MICHAEL LEHR • *Saarland University, Biopharmaceutics and Pharmaceutical Technology, Saarbruecken, Germany*

RAJSHREE MEWANI • *Departments of Radiation Medicine and Biochemistry, Molecular and Cellular Biology, Georgetown University Medical Center, Washington, DC 20057, USA*

LENA MÄLER • *Center for Biomembrane Research, Department of Biochemistry and Biophysics, Stockholm University, SE-106 91 Stockholm, Sweden*

JOHN J. NEMUNAITIS • *Mary Crowley Medical Research Center, Dallas, TX, USA*

ANDREA NEUMEYER • *Saarland University, Biopharmaceutics and Pharmaceutical Technology, Saarbruecken, Germany*

JOSEPH T. NEWSOME • *Lombardi Comprehensive Cancer Center, and Division of Comparative Medicine, Georgetown University Medical Center, Washington, DC 20057, USA*

DAVID OWEN • *Medical Research Council, Laboratory of Molecular Biology, Hills Road, Cambridge CB2 2QH, UK*

JIN PEI • *Departments of Radiation Medicine and Biochemistry, Molecular and Cellular Biology, Georgetown University Medical Center, Washington, DC 20057, USA*

YOSEF ROSENBERG • *Department of Physiology and Pharmacology, Sackler Faculty of Medicine, Tel-Aviv University, Ramat Aviv, 69978 Tel-Aviv, Israel*

LUDWIG G. STRAUSS • *Medical PET Group, Clinical Cooperation Unit Nuclear Medicine, German Cancer Research Center, Heidelberg, Germany*

BIRGIT WAHL • *Saarland University, Biopharmaceutics and Pharmaceutical Technology, Saarbruecken, Germany*

DONNA WILLIAMS • *Medical Research Council, Laboratory of Molecular Biology, Hills Road, Cambridge CB2 2QH, UK*

ANDERS WITTRUP • *Department of Clinical Sciences, Section of Oncology, Lund University, Sweden*

JON WOLFF • *Departments of Pediatrics and Medical Genetics, Waisman Center, University of Wisconsin-Madison, Madison, WI 53705, USA*

JOHN S. VORHIES • *Mary Crowley Medical Research Center, Dallas, TX, USA*

CHUANBO ZHANG • *Departments of Radiation Medicine and Biochemistry, Molecular and Cellular Biology, Georgetown University Medical Center, Washington, D.C. 20057, USA*

GREGORY T. ZUGATES • *Department of Biological Engineering, Department of Chemical Engineering, and Center for Cancer Research, Massachusetts Institute of Technology, Cambridge, MA 02139, USA*

Color Plates

Color Plate 3 Particle deposition in the lungs. Particles larger than 10 μm are deposited in the mouth and throat (*yellow/orange area*) and are swallowed. Particles between 6 and 10 μm are deposited in the upper airways (*blue area*). Particles between 0.5 and 6 μm are within the respirable range and are deposited in the alveolar region (*pink area*). Image from: http://www.filterair.info/articles/article.cfm/ArticleID/ 36856F0C-747B-4E08-B730798D614269E9/Page/1 (*See* discussion on p. 166.)

1

Developments in Macromolecular Drug Delivery

Mattias Belting and Anders Wittrup

Summary

Macromolecular drugs hold great promise as novel therapeutics of several major disorders, such as cancer and cardiovascular disease. However, their use is limited by lack of efficient, safe, and specific delivery strategies. Successful development of such strategies requires interdisciplinary collaborations involving researchers with expertise on, e.g., polymer chemistry, cell biology, nanotechnology, systems biology, advanced imaging methods, and clinical medicine. This not only poses obvious challenges to the scientific community but also provides opportunities for the unexpected at the interface between different disciplines. This introductory chapter summarizes and gives references to studies on macromolecular delivery that should be of interest to a broad scientific audience involved in macromolecular drug synthesis as well as in vitro and in vivo drug delivery studies.

Key Words: Gene transfer; Gene silencing; Drug delivery vehicles; Endocytosis; Peptides; Antibodies; Cancer; Positron emission tomography.

1. Introduction to Macromolecular Drug Delivery

The principle of macromolecular drug therapeutics is to correct pathological phenotypes by intracellular delivery of foremost nucleic acids and proteins. The great advantage with, e.g., nucleic acid-based therapeutics over small molecule drugs is their ability to act sequence-specifically on well-defined molecular targets, such as mutated or aberrant genes over-expressed in cancer cells. Hence, adverse effects due to structural issues are less likely than with small molecule inhibitors. The inherent challenge with these strategies, however, is poor delivery efficiency. The plasma membrane constitutes the most apparent barrier to

From: *Methods in Molecular Biology, vol. 480: Macromolecular Drug Delivery*, Edited by: M. Belting
DOI 10.1007/978-1-59745-429-2_1, © Humana Press, a part of Springer Science+Business Media, LLC 2009

the cellular uptake of macromolecules, and the rate of membrane passage varies depending on the size and hydrophobicity of the molecule. The characteristics of the lipid bilayer efficiently prevent cellular entry of hydrophilic molecules, i.e., nucleic acids and charged peptides. In the in vivo situation, the gut and airway mucosa, the blood–brain barrier, and the skin represent highly complex barriers that need to be penetrated in a controlled manner. The design of macromolecular drug delivery vehicles thus requires an increased understanding of the basic pathways involved in transcellular and intercellular transports without inflicting on the integrity of the cell (*see* **Fig. 1** for a schematic summary of approaches and methodologies used in macromolecular drug delivery research).

Cells exhibit several fundamental processes for the exchange of macromolecules, e.g., endocytosis, macropinocytosis, and phagocytosis that can be capitalized on for the delivery of macromolecular drugs *(1)*. Viruses and other types of microbes have developed several strategies to invade host cells via these pathways, and insights from studies of these strategies should be of great value for researchers in the drug delivery field. The ideal drug delivery vehicle may thus be a nanoparticle armed with all the invasive properties of a virus particle, but that can be synthesized and administered in a controlled manner without the risks of adverse immunologic or oncogenic events. Alternatively, biophysical approaches to temporarily disrupt the plasma membrane barrier with concomitant preservation of cell viability may be successful, at least for local drug delivery applications.

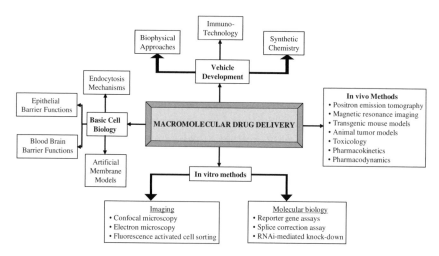

Fig. 1. Schematic overview of some of the major areas in macromolecular drug delivery research.

2. Vehicles for Macromolecular Delivery

2.1. Polymers and Cationic Lipids for Delivery of DNA and RNA In Vitro and In Vivo

The most studied molecule, so far, in the macromolecular drug delivery field is with no doubt nucleic acid, i.e., a polymer with repetitive negative charges. Transfection is a widely used non-viral strategy to deliver exogenous nucleic acid into eukaryotic cells, allowing detailed studies of the functional role of particular genes and their products. Naked DNA shows insignificant uptake over the cell membrane and is rapidly degraded by plasma nucleases, resulting in low transfection efficiency in vivo. Several different synthetic moieties, e.g., polyethylenimine (PEI), 2-diethylaminoether (DEAE)-dextran, artificial lipids, proteins, and peptides have been tested for their ability to condensate, protect, and deliver DNA into the intracellular compartment of various cells and tissues. Typically, they all have a positive net charge and are attracted not only to the negative charge of the target cell surface, but also to the negatively charged serum components. Over the last decades, the significant progress in non-viral gene delivery has failed to generate gene therapy protocols in the clinic. Nucleic acid delivery vehicles thus require further evolution in order to successfully introduce therapeutic genetic material into its appropriate destination in vivo.

Considering that the majority of malignant disorders are caused by acquired genetic abnormalities, it is not surprising that most pre-clinical development projects in the nucleic acid-based drug delivery field focus on tumor targeting. DNA vector-based short hairpin RNA (shRNA) as a means of effecting RNA interference (RNAi) is perhaps the most promising strategy for specific knockdown of mRNA to achieve targeted therapeutic effects in cancer disease. RNAi has become vital for gene silencing studies and has revolutionized experimental medical research as well as drug discovery. As is the case with DNA transfection studies, the critical and first step in RNAi is successful intracellular delivery, i.e., efficient macromolecular uptake with concomitant low cytotoxicity. Vorhies and Nemunaitis provide an excellent overview of leading candidates of biodegradable and non-biodegradable delivery vehicles for clinical usage *(2)*, and Gopalakrishnan and Wolff provide a summary of various strategies used for the delivery of DNA and small inhibitory RNA (siRNA) into mammalian cells in vitro *(3)*. They further provide a very useful description of several cationic lipid/polymer-based transfection reagents for the in vitro delivery of siRNA and DNA with emphasis on the key parameters that affect transfection efficacy. Some of the major challenges with various transfection protocols are discussed.

Green et al. describe the development of polymeric vectors, i.e., biodegradable poly(β-amino esters; PBAEs), for non-viral gene delivery *(4)*. These polymers have several properties that make them promising as gene delivery vehicles, i.e., they self-assemble with DNA into biodegradable, non-cytotoxic, positively charged nanoparticles. Indeed, PBAEs have been tested as gene delivery system in the treatment of prostate cancer. The authors further present a very useful strategy for high-throughput screening of new gene delivery biomaterials while varying important parameters such as polymer structure, DNA loading basis, and polymer-to-DNA weight ratio. Also, protocols for synthesis of PBAEs, particle self-assembly, and PBAE-mediated transfection are described.

It has been shown that the Raf-1 protein serine/threonine kinase is a druggable signaling molecule in cancer therapy *(5)*, and Kasid and co-workers have developed cationic liposomes for in vivo delivery of raf antisense oligonucleotides and raf siRNA into human tumor xenografts established in immunodeficient mice. The authors provide very useful protocols for the preparation of a modified cationic liposome/antisense oligonucleotide formulation as well as toxicology, pharmacokinetics, biodistribution, and anti-tumor efficacy studies in mice *(5)*.

2.2. Peptide-Based Delivery of Oligonucleotides and DNA Plasmids

Already in 1965, Ryser and Hancock provided evidence that histones and polyamino acids could greatly enhance albumin uptake by cultured tumor cells *(6)*. More recently, several polybasic peptides (so-called "protein transduction domains," PTDs or "cell-penetrating peptides," CPPs) have been shown to efficiently mediate uptake of nucleic acids, bioactive peptides, phage particles, and liposomes into a wide variety of mammalian cells. The initially proposed ability of CPPs to penetrate plasma membranes via a temperature-independent, non-endocytotic pathway was later shown to be a fixation artifact, and it is currently widely accepted that CPP-mediated macromolecular delivery follows energy-dependent endocytotic pathways that in most cases depend on the expression of cell-surface heparan sulfate proteoglycans (HSPGs) *(7)*.

The majority of human genes undergo alternative mRNA splicing and several acquired diseases (e.g., cancer) or genetic diseases (e.g., β-thalassemia and Duchenne muscular dystrophy) can potentially be treated through control of splicing by synthetic oligonucleotides (ON). Antisense ON, ribozymes, siRNA, microRNA (miRNA), triple-helix forming ON or aptamers control gene expression through specific interactions with RNA, DNA or even proteins. Current research aims at improving the metabolic stability of synthetic ON, and their selectivity and affinity during target recognition. Importantly, inefficient nuclear delivery of the correcting ON remains a major issue, which has been addressed by Gait and Lebleu, and co-workers *(8)*. They have developed procedures for the conjugation of CPPs with steric-block neutral DNA mimics, which may offer sequence-specific interference with pre-mRNA splicing with applications both in experimental models and as therapeutics. Moreover, the authors have established detailed protocols of how to monitor cellular uptake of CPP–ON conjugates, and to evaluate their biological activity in a splicing correction assay *(8)*.

CPP-mediated delivery of nucleic acids involves strategies based on chemical conjugation, as described above, as well as direct complex formation, based on the electrostatic interaction between negatively charged nucleic acids and polybasic CPPs. While there is relatively broad consensus on the notion that most CPP–nucleic acid complexes enter cells through endocytosis, it is still a matter of controversy as in which specific endocytotic pathway/s is involved. The most well-known CPP, the HIV-1 transactivator of transcription transduction domain (Tat) has, for example, been suggested to enter cells through clathrin-mediated endocytosis, caveolar endocytosis as well as macropinocytosis, partly depending on the size of internalized particles. The confusion regarding the mechanisms of CPP-mediated macromolecular delivery thus requires detailed studies that provide more in depth understanding of the endocytotic pathway involved under different conditions.

Wittrup and Belting have in several studies on CPP-mediated DNA delivery established protocols for fluorescence assisted cell sorting (FACS) analysis to obtain reliable, quantitative data on CPP–DNA complex uptake in cultured cancer cells *(9)*. Also, procedures for co-localization studies with known markers of various endocytotic pathways using confocal microscopy are described as well as the expression of dominant negative dynamin (GTPase deficient dynamin-2) to evaluate the dynamin dependence of the uptake mechanism. The use of various drugs commonly used to disrupt endocytosis is discussed, especially with regard to their limited specificity *(9)*.

2.3. Antibody-Based Targeting of Macromolecular Drugs

The use of polymeric vehicles and CPPs for macromolecular delivery is relatively safe and efficient, but how about their targeting specificity? As indicated above, most of these strategies are based on the electrostatic attraction between the positively charged cargo particle and the negatively charged surface of target cells. The polyanionic envelope surrounding virtually any mammalian cell is mainly provided by proteoglycans (PGs), i.e., proteins conjugated with linear polysaccharides substituted by sulfate groups at various positions *(7)*. Although PGs may be essential as a general portal of entry of macromolecules, and even may be expressed in a cell-specific manner, drug delivery vehicles must carry additional moieties that endow them with target specificity in vivo. Many attempts have been made in this direction, including the addition of the Arg-Gly-Asp (RGD) tripeptide, which recognizes integrins, and galactose that binds to specific membrane lectins present on hepatocytes. Other ligands that have been tested include transferrin, low density lipoprotein (LDL), asialooro-somucoid, epidermal growth factor (EGF), and folate. The most promising strategy, however, may be the use of antibodies directed at cell-surface antigens. The inherent features of monoclonal antibodies (mAbs), i.e., high specificity and generally low toxicity, should make them ideal for cell or tissue specific targeting. Upon antigen binding, the mAbs could induce intracellular signaling events that stimulate cytoskeleton rearrangement, membrane invagination, and eventually internalization into the target cells by endocytosis. Fransson and Borrebaeck have successfully established protocols for the retrieval of internalizing antibodies from phage display libraries, an approach that should be of great interest to researchers in the drug delivery field *(10)*. Phage display is a well-established technique for the selection of protein, e.g., antibodies, or peptide binders to various substrates. In most of the cases, the antigen is a polypeptide immobilized onto some kind of solid support with the risks that the antibody may only recognize its antigen in an artificial environment. To ensure the native configuration of the antigen in its original cellular context, phage selections are performed under more physiological conditions, where bound phages are allowed to internalize into the target cells. Fransson and Borreback also present a procedure for the identification of target antigens from intact cells by immunoprecipitation and MALDI-TOF analysis. Furthermore, an indirect immunotoxin cytotoxicity protocol is provided to assess the macromolecular delivery potential of selected antibody clones *(10)*.

3. Macromolecular Delivery and Endosomal Escape by Electroendocytosis

The net uptake of a macromolecular drug is not necessarily the most critical end-point as the therapeutic effect is highly dependent on its intracellular distribution. In fact, the major hurdle to efficient macromolecular drug delivery is not the initial steps of endocytosis, which *N.B.* does not involve membrane passage, but instead the escape from the endolysosomal compartment that requires actual membrane penetration. Viruses have developed several endosomal escape mechanisms, whereas non-viral gene transfer-vectors are considerably less sophisticated. Cationic lipids are thought to provoke endosomal membrane disruption through lipid exchange where anionic lipids from the cytoplasm-facing lipid layer flip-flop to the lumenal layer. The anionic lipids could then change places with the cationic lipids, resulting in the release of internalized DNA into the cytoplasm. As another example, PEI is supposed to escape through the so-called "proton sponge mechanism," i.e., a buffering mechanism where protons are pumped into the endosome, which subsequently leads to osmotic swelling and endosomal rupture. In all cases, these mechanisms are relatively inefficient and a major limitation to successful macromolecular delivery into, e.g., the cytoplasm or the nucleus.

A typical biological membrane is a complex structure composed primarily of lipids and proteins. Artificial biomembrane mimetic model systems can be used to characterize ligand–membrane interactions using a wide range of methods. These include monitoring of peptide-induced leakage as well as studies of the translocation of, e.g., CPPs across the membranes. Gräslund and co-workers have established several membrane model systems to investigate peptide–membrane interactions, and they describe methods for preparation of various membrane mimetic media, e.g., large unilamellar vesicles, micelles, and two-component bilayered micelles (bicelles), the latter of which have been used successfully to investigate the effect of bioactive peptides on lipid order and dynamics *(11)*.

Several mechanical or physical methods to deliver macromolecules across membranes have been described, e.g., ballistic bombardment of cells by micro particles coated with DNA, ultrasound, and electroporation where cells are exposed to high electric fields for micro- to milliseconds, resulting in transient membrane permeabilization thus allowing the diffusion of molecules across the membrane. Interestingly, Korenstein and co-workers have developed a system based on electroendocytosis for efficient incorporation of macromolecules into living cells *(12)*. The exposure of cells to a low electric field results in enhanced adsorption of macromolecules onto the target cell surface and increased endocytosis. Enhanced uptake of macromolecules is thus attributed both to the direct stimulation of different endocytic pathways as well as to the indirect effect mediated through the increase of the adsorption of the macromolecules onto the exposed cells. The method has found an application in the treatment of different metastatic tumor models, and protocols for applying electroendocytosis in macromolecular drug delivery are described *(12)*.

4. Macromolecular Transport Across Biological Barriers

Epithelial tissues, most importantly the gastrointestinal mucosa, skin, and the alveolar tract of the lung, are the primary barriers against uncontrolled uptake of a variety of potentially toxic substances. Accordingly, the development of strategies for controlled delivery of macromolecules across these biological barriers is crucial in macromolecular drug design. As intestine and lung epithelium only consist of single cell layers, they may be considered as best suited for efficient and controlled systemic delivery of macromolecular drugs. A powerful tool in this context is in vitro cell culture models imitating the more complex in vivo situation under controlled conditions. As nanoparticles are able to interfere with the biological barrier function of epithelia, their application as permeability enhancers for low permeable drugs is of special interest. Lehr and co-workers have established very useful procedures using polyethylenimide as model enhancer mimicking toxicological effects and altered barrier function in an epithelial in vitro model, which are described in detail *(13)*. They further provide protocols for cell viability and cytotoxicity assays to evaluate effects of macromolecular delivery enhancers.

For clinical purposes, systemic administration of macromolecules through the pulmonary route is especially attractive, since it offers a large, highly vascularized area that is available non-invasively. In addition, protein and peptide degradation is minimized due to the relatively low proteolytic activity found in the alveolar environment; and drugs absorbed through the lung avoid first-pass metabolism. Indeed, the first inhaled drug for the treatment of a systemic disease was recently approved (EXUBERA®), i.e., an inhaled insulin formulation. Several particle engineering approaches have been developed to produce dry powders with good bioavailability in the alveolar space. The principles and procedures of some of these methods, such as spray drying from solution, the formation of drug-containing liposomes, and the controlled crystallization of particles are reviewed and described by Kraft and Grant *(14)*.

The blood–brain barrier (BBB) represents a non-epithelial barrier of high clinical relevance. It prevents the passage of most macromolecules from the circulation into the brain parenchyma, which poses serious limitations in the treatment of central nervous system (CNS) malignancies. The BBB is constituted by the endothelial cells of the cerebral capillaries that are closely attached via tight junctions. In addition, the basal lamina provides a physiological barrier that actively removes undesirable molecules from the brain. Interestingly, non-invasive, transient, and magnetic resonance-guided BBB disruption can be accomplished by using focused ultrasound exposure with intravascular injection of pre-formed microbubbles. A few years ago, Hynynen and co-workers were able to demonstrate non-invasive and reversible disruption of the BBB at targeted locations using focused ultrasound in conjunction with an ultrasound contrast agent. Hynynen provides a detailed description of the method for magnetic resonance imaging (MRI)-guided focal BBB disruption in animals that should be of interest to anyone involved in CNS-targeted macromolecular drug delivery *(15)*.

5. Macromolecular Delivery for In Vivo Molecular Diagnostics

The use of nanoparticles and other macromolecular delivery vehicles in clinical medicine is not only limited to therapy but also offers novel diagnostic tools. Perhaps, the most promising methodology in this direction is positron emission tomography (PET), which has the capability of enhancing the specificity of diagnostics by the use of dedicated radiopharmaceuticals. PET has already been implemented, as a standard procedure for the evaluation of pathophysiological processes, like cellular metabolism, tumor perfusion, and expression of receptors in cancer. As such, it provides important information on primary tumor volume and the extent of dissemination prior to start of therapy as well as the response to given therapy. The most commonly used radiotracer for PET examinations, so far, is ^{18}F-fluorodeoxyglucose (FDG), i.e., a glucose analogue which reflects tumor cell viability based on the increased glucose uptake and glycolysis in malignancies. More recently, peptide-based tracers that are specific for cancer cell membrane receptors have attained major interest, as they provide specific molecular biological information non-invasively. Strauss and co-workers provide an excellent summary of current and future PET applications in cancer disease as well as step-by-step descriptions of tracer production, PET data acquisition, and dynamic data evaluation *(16)*.

6. Conclusions and Future Perspectives

The implementation of macromolecular drugs in the clinic is hampered by a limited understanding of the basic mechanisms of macromolecular membrane transport. Basic cell biological studies as well as in vivo experimental studies are required to provide a better understanding of how macromolecular complexes are processed by mammalian cells and tissues. Screening programs for novel synthetic vehicles of macromolecular drugs in various biological systems is another obvious focus of future studies. The "nanomedicine" hype has served an important role in stimulating chemists, molecular biologists, and clinicians to interdisciplinary collaborations on the design of the ideal delivery vehicle. Once available, macromolecular drugs will revolutionize the treatment of various diseases as well as provide novel diagnostic tools for the benefit of the patient.

References

1. Conner, S.D., and Schmid, S.L. (2003) Regulated portals of entry into the cell. *Nature* **422**, 37–44.
2. Vorhies, J.S., and Nemunaitis, J.J. (2008) Synthetic vs. natural/biodegradable polymers for delivery of shRNA based cancer therapies, in Macromolecular drug delivery (Belting, M., ed.), Humana Press, pp. 11–29.
3. Gopalakrishnan, B., and Wolff, J. (2008) siRNA and DNA transfer to cultured cells, in Macromolecular drug delivery (Belting, M., ed.), Humana Press, pp. 31–52.
4. Green, J.J., Zugates, G.T., Langer, R., and Anderson, D.G. (2008) Poly (β-amino esters): Procedures for synthesis and gene delivery, in Macromolecular drug delivery (Belting, M., ed.), Humana Press, pp. 53–63.
5. Zhang, C., Newsome, J.T., Mewani, R., Pei, J., Gokhale, P.C., and Kasid, U.N. (2008) Systemic delivery and preclinical evaluation of nanoparticles containing

antisense oligonucleotides and siRNAs, in Macromolecular drug delivery (Belting, M., ed.), Humana Press, pp. 65–83.

6. Ryser, H.J., and Hancock, R. (1965) Histones and basic polyamino acids stimulate the uptake of albumin by tumor cells in culture. *Science* **150**, 501.

7. Belting, M. (2003) Heparan sulfate proteoglycan as a plasma membrane carrier. *Trends Biochem. Sci.* **28**, 145–151.

8. Abes, S., Ivanova, G.D., Abes, R., Arzumanov, A.A., Williams, D., Owen, D., Lebleu, B., and Gait, M.J. (2008) Peptide-based delivery of steric-block PNA oligonucleotides, in Macromolecular drug delivery (Belting, M., ed.), Humana Press, pp. 85–99.

9. Wittrup, A., and Belting, M. (2008) Characterizing peptide mediated DNA internalization in human cancer cells, in Macromolecular drug delivery (Belting, M., ed.), Humana Press, pp. 101–112.

10. Fransson, J., and Borrebaeck, C.A.K. (2008) Selection and characterization of antibodies from phage display libraries against internalizing membrane antigens, in Macromolecular drug delivery (Belting, M., ed.), Humana Press, pp. 113–127.

11. Mäler, L., and Gräslund, A. (2008) Artificial membrane models for the study of macromolecular delivery, in Macromolecular drug delivery (Belting, M., ed.), Humana Press, pp. 129–139.

12. Barbul, A., Antov, Y., Rosenberg, Y., and Korenstein, R. (2008) Enhanced delivery of macromolecules into cells by electroendocytosis, in Macromolecular drug delivery (Belting, M., ed.), Humana Press, pp. 141–150.

13. Daum, N., Neumeyer, A., Wahl, B., Bur, M., and Lehr, C.-M. (2008) *In vitro* systems for studying epithelial transport of macromolecules, in Macromolecular drug delivery (Belting, M., ed.), Humana Press, pp. 151–164.

14. Kelly, K.S., and Grant, M. (2008) Preparation of macromolecule-containing dry powders for pulmonary delivery, in Macromolecular drug delivery (Belting, M., ed.), Humana Press, pp. 165–174.

15. Hynynen, K. (2008) Macromolecular delivery across the blood-brain barrier, in Macromolecular drug delivery (Belting, M., ed.), Humana Press, pp. 175–185.

16. Ludwig, G., Strauss, L.G., and Dimitrakopoulou-Strauss, A. (2008) Positron emission tomography (PET) and macromolecular delivery in vivo, in Macromolecular drug delivery (Belting, M., ed.), Humana Press, pp. 187–198.

2

Synthetic vs. Natural/Biodegradable Polymers for Delivery of shRNA-Based Cancer Therapies

John S. Vorhies and John J. Nemunaitis

Summary

DNA vector-based short hairpin RNA (shRNA) as a means of effecting RNA interference (RNAi) is a promising mechanism for the precise disruption of gene expression to achieve a therapeutic effect. The clinical usage of shRNA therapeutics in cancer is limited by obstacles related to effective delivery into the nuclei of target cancer cells. Significant pre-clinical data have been amassed about biodegradable and non-biodegradable polymeric delivery vehicles that are relevant for shRNA delivery into humans. Here, we will review some leading candidates for clinical usage with a focus on studies relating to their potential for usage in cancer shRNA therapeutics and discuss some of the advantages and disadvantages of using biodegradable and non-biodegradable delivery vehicles.

Key Words: shRNA; RNAi; Biodegradable polymer; Natural polymer; Synthetic polymer; Non-viral gene delivery; Drug delivery; Nanoparticle; Microparticle; Tumor targeting.

1. Introduction

The development of delivery vehicles is a crucial step toward the goal of safe and effective systemically administered RNA interference (RNAi)-based therapeutics. Unmodified nucleic acids are rapidly degraded in the bloodstream, therefore, an appropriate delivery vehicle must be developed in order to encapsulate nucleic acids, protecting them from host defenses and facilitating their entry into target cells. Great advances have been made in the development of polymeric delivery of drugs through modulations in their bioavailability, safety, and

From: *Methods in Molecular Biology, vol. 480: Macromolecular Drug Delivery*, Edited by: M. Belting
DOI 10.1007/978-1-59745-429-2_2, © Humana Press, a part of Springer Science+Business Media, LLC 2009

serum half-life. Many polymers that were originally developed to deliver small molecule drugs into human tissues can be adopted and modified for use with nucleic acid-based therapeutics such as RNAi. Polymers offer many advantages over lipid and viral delivery systems such as variable size, ease of production, and ease of conjugation to targeting moieties. The polymeric class of drug delivery vehicles is divided into polymers of natural origin that are biodegradable and polymers that are synthetic and not readily degradable in the body.

2. RNA Interference

(RNAi) is an endogenous post-transcriptional gene regulation phenomenon which can be harnessed to effect gene silencing in vitro and in vivo. RNAi is mediated by RNA duplexes of ~19 bp that have complete or almost complete complementarity to a target mRNA transcript. These small RNA duplexes interact with several protein complexes in the cytoplasm that mediate the binding of the duplex to the target mRNA and cause either direct inhibition of translation or a sequence specific enzymatic cleavage of target mRNA. Both of these pathways result in inhibition of gene expression. In naturally occurring RNAi, small nucleic acid duplexes with loops called pre-microRNAs (miRNAs) are transcribed in the nucleus and then transported into the cytoplasm through the nuclear membrane protein Exportin V. Once in the cytoplasm they are modified and then complexed with the RNA-induced silencing complex (RISC) inhibiting translation of the target mRNA (*1*).

The RNAi effect can be induced to interfere with the expression of a specific gene by the introduction of synthetic short interfering RNAs (siRNAs) into the cytoplasm or by the introduction of a DNA vector into the nucleus, which takes advantage of host transcription machinery to generate the RNAi effecting molecule (*2*). DNA vector-mediated RNAi is attractive when compared to the direct introduction of siRNA oligomers for many reasons including increased potential for cell type specificity through the use of selective promoters and the potential for enhanced duration of effect (*3,4*). In **Caenorhabditis** *elegans*, the model organism in which RNAi was originally studied, cytoplasmic RNA-dependent RNA polymerases replicate exogenous siRNA when it is introduced into the cytoplasm. Mammalian cells lack this RNA polymerase system, so the RNAi effect diminishes as the target cells divide (*5*). This poses a problem for RNAi therapy of cancer when in many cases the target cells are highly proliferative and thus rapidly dividing. DNA vector-based RNAi can prolong the interfering effect in human cells by exploiting the machinery already in place to transcribe and process miRNA.

The most promising method of accomplishing DNA vector-mediated RNAi has proven to be the expression of an RNA duplex with an end loop called short hairpin RNA (shRNA), from a single promoter. These shRNAs are then exported from the nucleus as if they were pre-miRNAs and are processed into siRNAs by Dicer or a functionally homologous double strand RNase *(6)*. This arrangement is analogous to the way in which endogenous miRNAs are encoded and transcribed. Because of overlaps with the natural miRNA system and competition for the same processing and transport proteins the total dose of shRNA must be monitored, so as not to saturate the system. Dose-dependent liver toxicity in mice has been observed with shRNA and is believed to be related to a bottleneck effect at the nuclear export step by overloading the Exportin V system *(7)*.

3. Considerations in shRNA Delivery

shRNA delivery has more in common with traditional gene therapy than it does with delivery of siRNA or antisense oligomers because the construct must eventually arrive in the nucleus of the target cells. Viral delivery has traditionally been the standard for in vivo shRNA, but concerns over safety and immunogenicity are driving interest in the development of safe, non-viral mechanisms for the mediation of vector-based RNAi *(8–10)*. For the purposes of human cancer therapy, a delivery vector must transport the genetic material to the tumor site(s) when given by local injection or by systemic administration. A good delivery vehicle suitable for systemic cancer therapy should be stable while in the bloodstream but be able to release its cargo efficiently and specifically inside the tumor cells. Vehicle size must be such that the complex can extravasate and penetrate the target tissue and target cells, and also avoid rapid renal clearance. The vehicle must not be overly immunogenic and must minimize nonspecific interactions. An optimal vehicle would also have the potential to bind targeting moieties. Here, we will briefly review key players in two categories of polymeric non-viral delivery vehicle systems: synthetic polymers and natural/biodegradable polymers (*see* **Table 1**).

3.1. Polymeric shRNA Delivery

Polymers for delivery of shRNA can be divided into two main categories; those of synthetic origins and those of natural origins that are biodegradable. Most synthetic polymers for drug or gene delivery have a backbone that is composed of carbon–carbon or amide bonds that are not easily degraded in physiological solutions. These polymers usually have high-transfection efficiency. Synthetic polymers reviewed here that show promise as potential shRNA delivery vehicles include dendrimers, polyethylenimine (PEI), and poly-L-lysine (PLL). Polymers of natural bases that are more easily degraded and excreted by the body are also good candidates for shRNA delivery. Natural polymers

Table 1
Key Players in Two Categories of Polymeric Non-viral Delivery Vehicle
Systems: Synthetic Polymers and Natural/Biodegradable Polymers

Vehicle	Description	References
Synthetic polymers		
Dendrimers	Highly branched polymer of well-defined size and charge	*(38–40)*
PEI (polyethylenimine)	Can be used in branched or linear form. Good potential for pulmonary therapy. Commonly used as copolymer component. Cationic density must be shielded	*(46–48)*
Poly-L-lysine (PLL)	An amino acid polymer with many positively charged amino groups. Its cytotoxicity can be mitigated by shielding	*(56)*
Natural/biodegradable polymers		
Collagen	Has been modified for extended release of genetic material	*(59)*
Gelatin	Derived from denatured collagen, its isoelectric point can be manipulated during formation to control charge. It has been modified for extended release of genetic material	*(61–63)*
Chitosan	An amino polysaccharide made by deacetylation of chitin. Its mucoadhesive properties make it ideal for oral and nasal administration	*(71–73)*
Cyclodextrins	Cyclic oligomers of glucose with a central hydrophobic cavity and a hydrophilic exterior	*(74,75)*
Poly(lactide-*co*-glycolide) (PLG)	A totally synthetic absorbable polymer with an extensive history of safe use in humans	*(76–79)*

reviewed here that show promise as potential shRNA delivery vehicles include collagen, gelatin, chitosan, alginate, poly(lactide-*co*-glycolide) (PLG) and their modified derivatives *(11–13)*.

Size and charge are the two properties of polymeric delivery systems that can be readily modified causing significant effect on in vivo performance. Most polymeric delivery vehicles are positively charged polymers that bind to DNA through charge interactions, encapsulating it, protecting it from host immune responses, and increasing its binding affinity to negatively charged cell membranes *(14)*. The complex of polymer and DNA is termed as polyplex. Cationic polymers are thought to effect compaction of the plasmid DNA through electrostatic interactions with the negatively charged phosphate groups, resulting in enhanced gene transfer activity *(15)*. Though polymer charge initially aids in DNA encapsulation and protection, too much charge can result in immune activation. Infusion of strongly charged particles can cause complement and immune activation, while neutral particles are less likely to be taken up by phagocytosis *(16,17)*. Particles with densely charged surfaces must be masked in order to avoid cytotoxicity as well as unwanted interactions with serum proteins, erythrocytes, and the reticuloendothelial system (RES). Charge shielding can be accomplished in most of the cases by complexing the particle with polyethylene glycol (PEG) or another similar compound *(18)*. Complexation with these molecules and modulation of charge increase circulation time, enhance the safety profile of the vehicle, and can also affect the vehicle's distribution within tissues *(19)*.

Charge can also aid in cellular penetration. Cationic nanoparticles typically gain entry into the cell by means of endocytosis. The positively charged particles readily attach to the external cell membrane's negatively charged glycocalyx and are subsequently internalized through various endocytotic mechanisms. Once the particle is endocytosed, the presence of multiple acid moieties with low pKa values facilitates endosomal destabilization and escape by means of the proton sponge effect; protons are continually pumped into the endosome and buffered by the polymer until the endosome becomes unstable due to the increasing ionic gradient *(20)*.

The size of the polyplex is also crucial to its function. The threshold for first-pass elimination by the kidneys is approximately 10 nm in diameter defining a rough lower size limit for nanoparticles *(21)*. Upper size limits are more difficult to establish as they depend on a variety of factors that are variable within tumors including penetration of capillary endothelium, diffusion rates in tumor interstitium and intracellular spaces *(22)*. Macromolecular complexes preferentially accumulate in tumors through the enhanced permeability and retention (EPR) effect *(23)*. Ideally, a nanoparticle would be in a size window such that it could take advantage of the EPR effect. The size of the polyplex can be readily modified during complexation by altering the DNA to polymer ratio *(24)*.

3.2. Targeting for Systemic Delivery

Systemic administration of an shRNA containing delivery vehicle in a clinical setting involves many more complications than in vitro transfection. Many polymeric delivery vehicles that are effective for shRNA delivery in vitro must be modified in order to be suitable for systemic delivery in the in vivo and clinical settings. The systemic biodistribution, pharmacokinetics, and pharmacodynamics of a polymer-delivered shRNA therapeutic can be optimized for the target by careful selection of the polymer and appropriate chemical modification of the polymer such as conjugation with aptamers, monoclonal antibodies or small molecules directed toward surface antigens on target.

Tumor cellular uptake at the target site can be further enhanced by promoting receptor-mediated endocytosis *(25)*. Polymeric-based delivery vehicles can be complexed with monoclonal antibodies for recognized cell surface markers as well as peptides, small molecule ligands, and aptamers directed to the target tumor cells.

Peptides, proteins, and monoclonalpt antibodies also have potential for targeting non-viral delivery of shRNA. Transferrin *(26)* fibroblast growth factor *(27)*, epidermal growth factor *(28)*, and vascular endothelial growth factor (VEGF) *(29)* among others have been used to target overexpression of receptors on cancer cells.

Small molecule ligands such as folate have proven to be excellent tools for use in tumor targeting. Small molecule conjugated therapeutics generally display stronger binding affinities and have better pharmacokinetic properties than those conjugated to larger antibodies because of their size and lack of immunogenicity *(30)*. Galactose and *asialoorosomucoid* have also been bound to delivery vehicles in order to target glycopeptide receptors on hepatocytes *(31)*.

Oligonucleotide ligands for identified tumor cell surface markers can also be selected and amplified from libraries *(32,33)*. Aptamers, which are chemically synthesized nucleotide oligomers with specific binding affinities for antigens, have been used to target various nanoparticulate drug delivery systems to cancerous cells. Aptamers have many advantages over monoclonal antibodies as targeting agents because their synthesis is non-biological, they are more stable, and they are non-immunogenic *(34)*. Aptamers are being actively developed for various surface proteins that are overexpressed on cancer cells. An aptamer directed to Nucleolin is currently under Phase 1 and Phase 2 clinical testing *(35)*, while aptamers directed to prostate specific membrane antigen, Tenascin-C *(36)*, and MUC1 *(37)* have undergone in vivo testing as targeting moieties and primary therapeutics.

4. Synthetic Polymers

4.1. Dendrimers

First described in 1978, dendrimers represent a relatively new structural class of polymer that has shown good potential for use as an effector of vector-based RNAi. Dendrimers consist of a central core molecule of which stems multiple arms of branched polymer *(38)*. Successive branches are created using a stepwise synthesis, which allows particle size to be precisely controlled *(39)*. With each step of branch synthesis, termed a generation, the number of branches increases exponentially causing an increase in polymer density correlating with radial position from the molecule's core. This pattern creates a physically protected void within the macromolecule. This void has chemical properties that differ from those of the surface, and can be exploited to host nucleic acids, much like a micelle *(40)*. Cationic dendrimers form complexes with DNA based on electrostatic interactions *(41)*. The multiple surface moieties can be used to complex targeting agents as well as other polymers to modify molecular properties.

Dendrimers used for DNA delivery most commonly have positive net surface charges. The most well studied molecules for gene delivery are those based on an ethylene diamine or ammonia core with polyamidoamine (PAMAM) dendrites or those based on butylenediamine cores and polypropylenimine (PPI) dendrites. Solubility of PAMAM dendrimers can be enhanced by partial acetylation of the reactive amino groups and they can be conjugated with various targeting entities such as folic acid to target receptors on specific cancer cells *(42)*.

In one study, a heavily boronated fifth-generation PAMAM dendrimer was conjugated to thiol groups of VEGF in order to target tumor neovasculature with overexpressed VEGF receptors. The resultant complex was tagged with a near-IR Cy5 dye to allow for near-IR fluorescent image monitoring of distribution in vivo. Following i.v. administration of the modified dendrimer in the murine xenograft 4T1 metastatic breast cancer model selective accumulation in tumors of VEGF bearing dendrimers was noted in comparison to naked controls particularly at the tumor periphery where angiogenesis was most active *(29)*. In another study, an HER2 specific monoclonal antibody was conjugated to a fifth-generation PAMAM dendrimers and fluorescently labeled. In a murine model, the conjugate showed increased binding to HER2-expressing subcutaneous tumor xenografts *(43)*. PAMAM dendrimers are being specifically evaluated as potential carriers of shRNA *(44)*.

PPI dendrimers have also demonstrated effectiveness as gene delivery vehicles in vivo *(24)*. In one study, commercially available PPI dendrimers were modified to enhance internal cationic quaternization by adding PEG-like acetyl groups or glycol gallate at the exterior primary amines and by adding methyl iodide or methyl chloride at the interior tertiary amines. Administration of this modified dendrimer intravenously complexed with single stranded DNA (ssDNA) to nude mice resulted in high nuclear uptake as confirmed by co-localization studies *(45)*. Both PPI and PAMAM dendrimers have shown promise as effective shRNA delivery vehicles for systemic administration and targeting of solid tumors.

4.2. Polyethylenimine (PEI)

The cationic polymer, PEI, has been used as a gene carrier for more than a decade and thus its safety profiles and transfection efficiencies under many conditions are well constrained. PEI consists of linear or branched alkyl chains interspersed with amines. Most of the amines are secondary but naturally the relative amount of primary, secondary, and tertiary amines varies with the degree of polymer branching.

The densely cationic nature of pure PEI can cause the polymer to have a significant cytotoxic effect. Approximately two-thirds of the amines are protonated at physiological pH *(46)* and the remainder of the amines contribute to the charge and buffering effects of the particle *(20)*. Linear PEI usually shows higher transfection efficiencies and lower cytotoxicities as compared to branched PEI *(47)*. In linear PEI, both cytotoxicity and transfection efficiency have been correlated with cationic density and molecular weight of the polymer *(48)*. Grafting PEI with PEG has been demonstrated to abrogate its cytotoxicity while having no significant effect on transfection efficiency *(49)*.

Conjugation of a DNA/PEI or DNA/PEI–PEG polyplex to targeting moieties such as antibodies to surface antigens like HER2 for breast cancer and OA3 for ovarian cancer has been shown to increase the transfection efficiency in vitro *(50,51)*. PEI can also be lined with amphipathic peptides in order to facilitate cell entry in vivo. PEI coated with the KALA (WEAK-LAKALAKALAKHLAKALAKALKACEA) peptide has been shown to have higher transfection efficiency in a murine model than plain PEI or a commercial liposome *(52)*.

PEI has been used as a delivery vehicle in several in vivo administrations of shRNA. In a murine xenograft model of Ewing's sarcoma, TC71 cell xenograft tissue transfected with PEI conjugated with a plasmid encoding siRNA for VEGF grew significantly less than the control tissue and had less vessel density as assessed by CD31 immunohistochemical analysis *(53)* (*see also* **Table 2**).

In a study of liver regeneration, plasmids encoding shRNA sequences specific to hepatocyte growth factor (HGF) and its receptor c-Met were complexed with linear PEI and injected into rats. shRNA treatment resulted in the suppression of c-Met and HGF mRNA and protein as well as moderate suppression of hepatocyte proliferation compared with that in mismatch and scramble controls *(54)*. In another experiment, linear PEI was used to deliver an shRNA construct targeting the thyroid hormone receptor $\alpha 1$ (TRα1) driven by a hybrid CMV-H1 promoter construct. The polyplexes were administered through stereotaxic injection into the lateral ventricles of the brains of newborn and adult mice where they strongly repressed CyclinD1 expression, which is a negatively regulated thyroid hormone target gene *(55)*.

Table 2
Polymeric Delivery Vehicle

Vehicle	Targeting	Promoter	Route	Disease model	Model system	Target site	Target mRNA	Result	Reference
Branched PEI	None	Pol-III U6	Intratumoral	Ewing's sarcoma (TC71)	Athymic nude mice	Subcutaneous	VEGF	Significantly inhibited tumor growth and increased survival time	*(53)*
Linear PEI	None	Pol-III H1	IV	Partial liver resection	Fisher rats	Liver	HGF	Suppression of mRNA and protein, moderate suppression of hepatocyte proliferation	*(54)*
Linear PEI	None	Pol-III H1	IV	Partial liver resection	Fisher rats	Liver	c-Met	Suppression of mRNA and protein, moderate suppression of hepatocyte proliferation	*(54)*
Linear PEI	None	Pol-III U6	Intratumoral	U87 human glioblastoma	SCID mice	Subcutaneous	VEGF	Efficient reduction of tumor vascularization	*(80)*

(Continued)

Table 2
(Continued)

Vehicle	Targeting	Promoter	Route	Disease model	Model system	Target site	Target mRNA	Result	Reference
Linear PEI	None	CMV-H1	Stereotaxic injection into lateral ventricles	Normal brain	Mice	Brain	TRα1	Inhibition of gene expression	(55)
Cationized gelatin nanoparticle	None	Pol-III H1	Intraureteral	Obstructive nephropathy	C57BL/6 mice	Kidney	TGF-β	Reduction in collagen content and fibrotic tissue in the kidney interstitium	(64, 81)
Cationized gelatin microparticles	None	Pol-III U6	Intratumoral	NRS-1 squamous cell carcinoma	C3H/He mice	Subcutaneous	VEGF	Suppression of tumor growth, prolonged knockdown, and decreased tumor microvascularity as compared to controls	(65)

CMV = Cytomegalovirus; EGFR = Epidermal growth factor receptor; HGF = Hepatocyte growth factor; PEG = Polyethylene glycol; PEI = Polyethylenimine; Pol = Polymerase; SCID = Severe combined immunodeficiency; TGF = TRα1; Endogenous thyroid hormone α1 transforming growth factor; VEGF = Vascular endothelial growth factor.

4.3. Poly-L-Lysine (PLL)

Lysine is a simple amino acid which has an R group that consists of an amino group at the end of a four carbon chain. The linear polypeptide PLL is able to effectively complex with DNA using its many positively charged amino groups, forming a stable polyplex. PLLs R amino group also allows for branching of the polymer to modulate structure and charge. Synthesis of branched PLL can even be controlled in a stepwise manner to create dendrimers *(56)*. PLL alone is highly cytotoxic, but charge shielding of PLL with PEG sufficiently mitigates the cytotoxicity to make it a useful in vivo gene delivery vehicle with potential for therapeutic delivery of shRNA. Much like PEI, PEG–PLL is capable of promoting the proton sponge effect and can be grafted with a fusogenic peptide such as KALA to enhance cell entry *(33)*.

PLL-based carriers have been used as vehicles for DNA vector-based RNAi in combination with a multifunctional envelope-type nanodevice. This combination complexed to a DNA plasmid encoding anti-luciferase siRNA demonstrated 96% inhibition of luciferase activity in an in vitro co-transfection study *(57)*.

5. Natural/Biodegradable Polymeric Delivery Systems

5.1. Collagen

Some modified collagen-based polymers such as atelocollagen have been investigated as potential gene deliver vehicles *(58)*. Methylated collagen, which is more stable and more condensed at physiologic pH than native collagen, has been studied in vivo as a gene delivery vehicle using luciferase reporter genes to monitor transfection efficiency. Methylation appears to increase stability while reducing in vivo transfection efficiency when compared to natural collagen *(59,60)*.

5.2. Gelatin

Gelatin, which is a simply denatured collagen, has shown promise in vitro and in vivo as an shRNA delivery vehicle. Cationized gelatin nanoparticles are relatively simple to produce when compared to synthetic polymers *(61)*. They have been shown to have an in vitro transfection efficiency of approximately one order of magnitude less than PEI, but show approximately four-fold less cytotoxicity *(62,63)*.

Cationized gelatin has been used in vivo to mediate vector-based RNAi in a murine model of obstructive nephropathy after intraureteral delivery. Administration of a plasmid encoding siRNA against the transforming growth factor-β (TGF-β) receptor gene resulted in a reduction in collagen content and fibrotic tissue in the kidney interstitium for up to ten days after administration *(64)*. Gelatin has proven to be an effective mediator of DNA vector-based RNAi in an NRS-1 squamous cell carcinoma murine xenograft model. In this study, a vector encoding siRNA targeted against VEGF mRNA was complexed to cationized gelatin microspheres and administered in vitro and in vivo. Effective knockdown was observed in vitro and administration resulted in suppressed tumor growth and reduced vascularity in vivo *(65)*.

5.3. Chitosan

Chitosan is one of the better studied natural cationic polymeric gene carriers. It is obtained by the deacetylation of chitin, which results in a biodegradable polysaccharide composed of two subunits, D-glucosamine and *N*-acetyl-D-glucosamine *(66)*. Chitosan has been extensively studied in the context of oligonucleotides and small molecule drug delivery *(67–69)*. Much the same as a synthetic cationic polymer, oligomeric or polymeric chitosan complexes readily with DNA through charge interaction and spontaneously forms microspheric or nanospheric particles through coacervative encapsulation *(70,71)*. Because of its mucoadhesive properties it has been particularly studied in experiments involving oral and nasal administration *(72)*. In vivo, oligomeric chitosan has demonstrated comparable transfection efficiency to PEI after murine lung administration *(73)*.

5.4. Cyclodextrin

Cyclodextrins are cyclic oligomers of glucose with an amphipathic structure that confers upon them unique properties that make them good candidates for therapeutic shRNA delivery. They have a central hydrophobic cavity and a hydrophilic exterior making them water soluble. Cyclodextrin can be made into cationic polymers by condensation of the monomers *(74)*. The resultant cationic polymer is capable of forming stable nanoparticle complexes with DNA that have low toxicity and transfection efficiencies similar to PEI and Lipofectamine *(75)*. Recent work has demonstrated that α-cyclodextrin can be conjugated to starburst® third generation PAMAM dendrimers to enhance their ability to deliver plasmids encoding shRNA to NIH3T cells in vitro. In this study, dendrimers conjugated with cyclodextrin showed higher transfection efficiency than dendrimers alone. When conjugated with plasmids encoding shRNA the addition of cyclodextrin to the dendrimers also enhanced the RNAi effect *(44)*.

5.5. Poly(lactide-co-glycolide) (PLG or PLGA)

PLG microparticles are leaders among biodegradable polymers because of their history of safe use in humans. PLG has been primarily explored as a delivery vehicle for intramuscular and intravenous administration of peptides such as luteinizing hormone releasing hormone agonist peptides such as Lupron Depot (TAP Pharmaceuticals), Zoladex (AstraZeneca) and proteins such as human growth hormone recombinant protein, Nutropin Depot (Genentech). PLG is also currently being investigated as a sole polymer and a copolymer component for the delivery of DNA *(76,77)*. DNA is adsorbed into the PLG microparticle with or without the aid of a cationic surfactant such as cetyltrimethylammonium bromide *(78)*. PLG is ideal for extended release preparations because when it is delivered via direct intramuscular injection it degrades slowly and releases its payload over an extended period of time.

In one study, PLG microspheres bound to DNA using a cryogenic double emulsion process and injected intramuscularly into mice were found to promote transgene expression for up to 174 days after injection, dependent upon microsphere mass. More recently nanoparticles were constructed from carboxy terminated, PEG modified PGL polymers. These particles were conjugated with an aptamer to the prostate specific membrane antigen and evaluated for in vivo biodistribution in an LNCaP (PSMA+) xenograft mouse model of prostate cancer. In this study, the aptamer caused a 3.77-fold increase in polymer concentration in the tumor tissue after retro-orbital injection *(79)*.

6. Synthetic Polymers vs. Natural/Biodegradable Polymers

Though less studied than synthetic polymer and lipid-based gene delivery systems, there have been significant advances in the field of natural polymers for gene delivery. Biodegradable polymers are less likely to accumulate within tissues than their synthetic counterparts. Synthetic polymers are not easily removed by normal clearance systems and can accumulate in tissues.

This primary difference between the two classes of delivery vehicles is quite significant to the pharmacokinetics, pharmacodynamics, and biodistribution of the polyplex.

Toxicity can result from shRNA expression in non-target tissue and accumulation of shRNA in the delivery vehicle in the liver or the kidneys. Transfection into non-cancerous cells can be especially deleterious in shRNA therapeutics because the RNAi target may be an essential transcript. Delivery vehicles that are resistant to degradation increase the serum half-life of the therapeutic and may increase its penetration in poorly perfused areas of tumor tissue. On the other hand, this also increases the risk of toxicity due to accumulation. Rapid polyplex degradation might be more likely to have a better safety profile but may decrease the concentration of the therapeutic in the target tissue with poor circulation.

7. Conclusions

Many excellent options exist for synthetic and natural polymer-mediated delivery of shRNA. At present, synthetic polymers are limited by safety concerns and natural, biodegradable polymers are limited by efficacy issues. Active research in the field is causing these limitations to become less meaningful, but at present the choice to use a synthetic or a natural polymer for shRNA delivery must be made by considering the nature and location of the tumor(s) to be treated as well as the type of transcript to be targeted in order to maximize safety and optimize effectiveness.

References

1. Matzke, M.A., and Birchler, J.A. (2005) RNAi-mediated pathways in the nucleus. *Nature Reviews* **6**:24–35.
2. Paddison, P.J., Caudy, A.A., Bernstein, E., Hannon, G.J., and Conklin, D.S. (2002) Short hairpin RNAs (shRNAs) induce sequence-specific silencing in mammalian cells. *Genes and Development* **16**:948–958.
3. Stein, U., Walther, W., Stege, A., Kaszubiak, A., Fichtner, I., and Lage, H. (2008) Complete in vivo reversal of the multidrug resistance phenotype by jet-injection of anti-MDR1 short hairpin RNA-encoding plasmid DNA. *Molecular Therapy* **16**(1):178–186.
4. Lage, H. (2005) Potential applications of RNA interference technology in the treatment of cancer. *Future Oncology* **1**:103–113.
5. Bantounas, I., Phylactou, L.A., and Uney, J.B. (2004) RNA interference and the use of small interfering RNA to study gene function in mammalian systems. *Journal of Molecular Endocrinology* **33**:545–557.
6. Wadhwa, R., Kaul, S.C., Miyagishi, M., and Taira, K. (2004) Vectors for RNA interference. *Current Opinion in Molecular Therapeutics* **6**:367–372.
7. Marsden, P.A. (2006) RNA interference as potential therapy–not so fast. *The New England Journal of Medicine* **355**:953–954.
8. Balter, M. (2000) Gene therapy on trial. *Science* **288**:951–957.
9. Check, E. (2002) A tragic setback. *Nature* **420**:116–118.
10. Nishikawa, M., and Hashida, M. (2002) Nonviral approaches satisfying various requirements for effective in vivo gene therapy. *Biological and Pharmaceutical Bulletin* **25**:275–283.
11. Dang, J.M., and Leong, K.W. (2006) Natural polymers for gene delivery and tissue engineering. *Advanced Drug Delivery Reviews* **58**:487–499.
12. Storrie, H., and Mooney, D.J. (2006) Sustained delivery of plasmid DNA from polymeric scaffolds for tissue engineering. *Advanced Drug Delivery Reviews* **58**:500–514.
13. Tiera, M.J., Winnik, F.O., and Fernandes, J.C. (2006) Synthetic and natural polycations for gene therapy: state of the art and new perspectives. *Current Gene Therapy* **6**:59–71.

14. Park, T.G., Jeong, J.H., and Kim, S.W. (2006) Current status of polymeric gene delivery systems. *Advanced Drug Delivery Reviews* **58**:467–486.
15. Schwartz, B., Ivanov, M.A., Pitard, B., Escriou, V., Rangara, R., Byk, G., Wils, P., Crouzet, J., and Scherman, D. (1999) Synthetic DNA-compacting peptides derived from human sequence enhance cationic lipid-mediated gene transfer in vitro and in vivo. *Gene Therapy* **6**:282–292.
16. Plank, C., Mechtler, K., Szoka Jr., F.C., and Wagner, E. (1996) Activation of the complement system by synthetic DNA complexes: a potential barrier for intravenous gene delivery. *Human Gene Therapy* **7**:1437–1446.
17. Chonn, A., Cullis, P.R., and Devine, D.V. (1991) The role of surface charge in the activation of the classical and alternative pathways of complement by liposomes. *Journal of Immunology* **146**:4234–4241.
18. Kircheis, R., Blessing, T., Brunner, S., Wightman, L., and Wagner, E. (2001) Tumor targeting with surface-shielded ligand–polycation DNA complexes. *Journal of Controlled Release* **72**:165–170.
19. Barron, L.G., Gagne, L., and Szoka Jr., F.C. (1999) Lipoplex-mediated gene delivery to the lung occurs within 60 minutes of intravenous administration. *Human Gene Therapy* **10**:1683–1694.
20. Boussif, O., Lezoualc'h, F., Zanta, M.A., Mergny, M.D., Scherman, D., Demeneix, B., and Behr, J.P. (1995) A versatile vector for gene and oligonucleotide transfer into cells in culture and in vivo: polyethylenimine. *Proceedings of the National Academy of Sciences of the United States of America* **92**:7297–7301.
21. Venturoli, D., and Rippe, B. (2005) Ficoll and dextran vs. globular proteins as probes for testing glomerular permselectivity: effects of molecular size, shape, charge, and deformability. *American Journal of Physiology Renal Physiology* **288**:F605–F613.
22. Dreher, M.R., Liu, W., Michelich, C.R., Dewhirst, M.W., Yuan, F., and Chilkoti, A. (2006) Tumor vascular permeability, accumulation, and penetration of macromolecular drug carriers. *Journal of the National Cancer Institute* **98**:335–344.
23. Greish, K. (2007) Enhanced permeability and retention of macromolecular drugs in solid tumors: a royal gate for targeted anticancer nanomedicines. *Journal of Drug Targeting* **15**:457–464.
24. Schatzlein, A.G., Zinselmeyer, B.H., Elouzi, A., Dufes, C., Chim, Y.T., Roberts, C.J., Davies, M.C., Munro, A. et al. (2005) Preferential liver gene expression with polypropylenimine dendrimers. *Journal of Controlled Release* **101**: 247–258.
25. Hughes, J.A., and Rao, G.A. (2005) Targeted polymers for gene delivery. *Expert Opinion on Drug Delivery* **2**:145–157.
26. Singh, M. (1999) Transferrin as a targeting ligand for liposomes and anticancer drugs. *Current Pharmaceutical Design* **5**:443–451.
27. Li, D., Wang, Q.Q., Tang, G.P., Huang, H.L., Shen, F.P., Li, J.Z., and Yu, H. (2006) Receptor-mediated gene delivery using polyethylenimine (PEI) coupled with polypeptides targeting FGF receptors on cells surface. *Journal of Zhejiang University Science B* **7**:906–911.
28. Liu, X., Tian, P., Yu, Y., Yao, M., Cao, X., and Gu, J. (2002) Enhanced antitumor effect of EGF R-targeted p21WAF-1 and GM-CSF gene transfer in the established murine hepatoma by peritumoral injection. *Cancer Gene Therapy* **9**: 100–108.
29. Backer, M.V., Gaynutdinov, T.I., Patel, V., Bandyopadhyaya, A.K., Thirumamagal, B.T., Tjarks, W., Barth, R.F., Claffey, K., and Backer, J.M. (2005) Vascular

endothelial growth factor selectively targets boronated dendrimers to tumor vasculature. *Molecular Cancer Therapeutics* **4**:1423–1429.

30. Hilgenbrink, A.R., and Low, P.S. (2005) Folate receptor-mediated drug targeting: from therapeutics to diagnostics. *Journal of Pharmaceutical Science* **94**: 2135–2146.

31. Anwer, K., Logan, M., Tagliaferri, F., Wadhwa, M., Monera, O., Tung, C.H., Chen, W., Leonard, P. et al. (2000) Synthetic glycopeptide-based delivery systems for systemic gene targeting to hepatocytes. *Pharmaceutical Research* **17**: 451–459.

32. Mori, T. (2004) Cancer-specific ligands identified from screening of peptide-display libraries. *Current Pharmaceutical Design* **10**:2335–2343.

33. Lee, J.F., Stovall, G.M., and Ellington, A.D. (2006) Aptamer therapeutics advance. *Current Opinion in Chemical Biology* **10**:282–289.

34. Nimjee, S.M., Rusconi, C.P., and Sullenger, B.A. (2005) Aptamers: an emerging class of therapeutics. *Annual Review of Medicine* **56**:555–583.

35. Ireson, C.R., and Kelland, L.R. (2006) Discovery and development of anticancer aptamers. *Molecular Cancer Therapeutics* **5**:2957–2962.

36. Hicke, B.J., Stephens, A.W., Gould, T., Chang, Y.F., Lynott, C.K., Heil, J., Borkowski, S., Hilger, C.S. et al. (2006) Tumor targeting by an aptamer. *Journal of Nuclear Medicine* **47**:668–678.

37. Perkins, A.C., and Missailidis, S. (2007) Radiolabelled aptamers for tumour imaging and therapy. *The Quarterly Journal of Nuclear Medicine and Molecular Imaging* **51**:292–296.

38. Dufes, C., Uchegbu, I.F., and Schatzlein, A.G. (2005) Dendrimers in gene delivery. *Advanced Drug Delivery Reviews* **57**:2177–2202.

39. Urdea, M.S., and Horn, T. (1993) Dendrimer development. *Science* **261**:534.

40. Alper, J. (1991) Rising chemical "stars" could play many roles. *Science* **251**: 1562–1564.

41. Bielinska, A.U., Chen, C., Johnson, J., and Baker Jr., J.R. (1999) DNA complexing with polyamidoamine dendrimers: implications for transfection. *Bioconjugate Chemistry* **10**:843–850.

42. Majoros, I.J., Thomas, T.P., Mehta, C.B., and Baker Jr., J.R. (2005) Poly (amidoamine) dendrimer-based multifunctional engineered nanodevice for cancer therapy. *Journal of Medicinal Chemistry* **48**:5892–5899.

43. Shukla, R., Thomas, T.P., Peters, J.L., Desai, A.M., Kukowska-Latallo, J., Patri, A.K., Kotlyar, A., and Baker Jr., J.R. (2006) HER2 specific tumor targeting with dendrimer conjugated anti-HER2 mAb. *Bioconjugate Chemistry* **17**: 1109–1115.

44. Tsutsumi, T., Hirayama, F., Uekama, K., and Arima, H. (2008) Potential use of polyamidoamine dendrimer/alpha-cyclodextrin conjugate (generation 3, G3) as a novel carrier for short hairpin RNA-expressing plasmid DNA. *Journal of Pharmaceutical Science* **97**(8):3022–3034.

45. Tack, F., Bakker, A., Maes, S., Dekeyser, N., Bruining, M., Elissen-Roman, C., Janicot, M., Brewster, M. et al. (2006) Modified poly(propylene imine) dendrimers as effective transfection agents for catalytic DNA enzymes (DNAzymes). *Journal of Drug Targeting* **14**:69–86.

46. Garnett, M.C. (1999) Gene-delivery systems using cationic polymers. *Critical Reviews in Therapeutic Drug Carrier System* **16**:147–207.

47. Wightman, L., Kircheis, R., Rossler, V., Carotta, S., Ruzicka, R., Kursa, M., and Wagner, E. (2001) Different behavior of branched and linear polyethylenimine for gene delivery in vitro and in vivo. *The Journal of Gene Medicine* **3**:362–372.
48. Jeong, J.H., Song, S.H., Lim, D.W., Lee, H., and Park, T.G. (2001) DNA transfection using linear poly(ethylenimine) prepared by controlled acid hydrolysis of poly(2-ethyl-2-oxazoline). *Journal of Controlled Release* **73**:391–399.
49. Petersen, H., Fechner, P.M., Martin, A.L., Kunath, K., Stolnik, S., Roberts, C.J., Fischer, D., Davies, M.C., and Kissel, T. (2002) Polyethylenimine-graft-poly(ethylene glycol) copolymers: influence of copolymer block structure on DNA complexation and biological activities as gene delivery system. *Bioconjugate Chemistry* **13**:845–854.
50. Chiu, S.J., Ueno, N.T., and Lee, R.J. (2004) Tumor-targeted gene delivery via anti-HER2 antibody (trastuzumab, Herceptin) conjugated polyethylenimine. *Journal of Controlled Release* **97**:357–369.
51. Merdan, T., Callahan, J., Petersen, H., Kunath, K., Bakowsky, U., Kopeckova, P., Kissel, T., and Kopecek, J. (2003) Pegylated polyethylenimine-Fab' antibody fragment conjugates for targeted gene delivery to human ovarian carcinoma cells. *Bioconjugate Chemistry* **14**:989–996.
52. Min, S.H., Lee, D.C., Lim, M.J., Park, H.S., Kim, D.M., Cho, C.W., Yoon, D.Y., and Yeom, Y.I. (2006) A composite gene delivery system consisting of polyethylenimine and an amphipathic peptide KALA. *The Journal of Gene Medicine* **8**:1425–34.
53. Guan, H., Zhou, Z., Wang, H., Jia, S.F., Liu, W., and Kleinerman, E.S. (2005) A small interfering RNA targeting vascular endothelial growth factor inhibits Ewing's sarcoma growth in a xenograft mouse model. *Clinical Cancer Research* **11**:2662–2669.
54. Paranjpe, S., Bowen, W.C., Bell, A.W., Nejak-Bowen, K., Luo, J.H., and Michalopoulos, G.K. (2007) Cell cycle effects resulting from inhibition of hepatocyte growth factor and its receptor c-Met in regenerating rat livers by RNA interference. *Hepatology (Baltimore, Md)* **45**:1471–1477.
55. Hassani, Z., Francois, J.C., Alfama, G., Dubois, G.M., Paris, M., Giovannangeli, C., and Demeneix, B.A. (2007) A hybrid CMV-H1 construct improves efficiency of PEI-delivered shRNA in the mouse brain. *Nucleic Acids Research* **35**:e65.
56. Yamagata, M., Kawano, T., Shiba, K., Mori, T., Katayama, Y., and Niidome, T. (2007) Structural advantage of dendritic poly(L-lysine) for gene delivery into cells. *Bioorganic and Medicinal Chemistry* **15**:526–532.
57. Moriguchi, R., Kogure, K., Akita, H., Futaki, S., Miyagishi, M., Taira, K., and Harashima, H. (2005) A multifunctional envelope-type nano device for novel gene delivery of siRNA plasmids. *International Journal of Pharmaceutics* **301**:277–285.
58. Sano, A., Maeda, M., Nagahara, S., Ochiya, T., Honma, K., Itoh, H., Miyata, T., and Fujioka, K. (2003) Atelocollagen for protein and gene delivery. *Advanced Drug Delivery Reviews* **55**:1651–1677.
59. Wang, J., Lee, I.L., Lim, W.S., Chia, S.M., Yu, H., Leong, K.W., and Mao, H.Q. (2004) Evaluation of collagen and methylated collagen as gene carriers. *International Journal of Pharmaceutics* **279**:115–126.
60. Cohen-Sacks, H., Elazar, V., Gao, J., Golomb, A., Adwan, H., Korchov, N., Levy, R.J., Berger, M.R., and Golomb, G. (2004) Delivery and expression

of pDNA embedded in collagen matrices. *Journal of Controlled Release* **95**: 309–320.

61. Coester, C.J., Langer, K., van Briesen, H., and Kreuter, J. (2000) Gelatin nanoparticles by two step desolvation – a new preparation method, surface modifications and cell uptake. *Journal of Microencapsulation* **17**:187–193.

62. Zwiorek, K., Kloeckner, J., Wagner, E., and Coester, C. (2005) Gelatin nanoparticles as a new and simple gene delivery system. *Journal of Pharmacy and Pharmaceutical Science* **7**:22–28.

63. Coester, C. (2003) Development of a new carrier system for oligonucleotides and plasmids based on gelatin nanoparticles. *New Drugs* **1**:14–17.

64. Kushibiki, T., Nagata-Nakajima, N., Sugai, M., Shimizu, A., and Tabata, Y. (2006) Enhanced anti-fibrotic activity of plasmid DNA expressing small interference RNA for TGF-beta type II receptor for a mouse model of obstructive nephropathy by cationized gelatin prepared from different amine compounds. *Journal of Controlled Release* **110**:610–617.

65. Matsumoto, G., Kushibiki, T., Kinoshita, Y., Lee, U., Omi, Y., Kubota, E., and Tabata, Y. (2006) Cationized gelatin delivery of a plasmid DNA expressing small interference RNA for VEGF inhibits murine squamous cell carcinoma. *Cancer Science* **97**:313–321.

66. Chandy, T., and Sharma, C.P. (1990) Chitosan – as a biomaterial. *Biomaterials, Artificial Cells, and Artificial Organs* **18**:1–24.

67. Howard, K.A., Rahbek, U.L., Liu, X., Damgaard, C.K., Glud, S.Z., Andersen, M.O., Hovgaard, M.B., Schmitz, A. et al. (2006) RNA interference in vitro and in vivo using a novel chitosan/siRNA nanoparticle system. *Molecular Therapy* **14**:476–484.

68. Katas, H., and Alpar, H.O. (2006) Development and characterisation of chitosan nanoparticles for siRNA delivery. *Journal of Controlled Release* **115**:216–225.

69. Agnihotri, S.A., Mallikarjuna, N.N., and Aminabhavi, T.M. (2004) Recent advances on chitosan-based micro- and nanoparticles in drug delivery. *Journal of Controlled Release* **100**:5–28.

70. Mansouri, S., Lavigne, P., Corsi, K., Benderdour, M., Beaumont, E., and Fernandes, J.C. (2004) Chitosan-DNA nanoparticles as non-viral vectors in gene therapy: strategies to improve transfection efficacy. *European Journal of Pharmaceutics and Biopharmaceutics* **57**:1–8.

71. Leong, K.W., Mao, H.Q., Truong-Le, V.L., Roy, K., Walsh, S.M., and August, J.T. (1998) DNA-polycation nanospheres as non-viral gene delivery vehicles. *Journal of Controlled Release* **53**:183–193.

72. Borchard, G. (2001) Chitosans for gene delivery. *Advanced Drug Delivery Reviews* **52**:145–150.

73. Koping-Hoggard, M., Tubulekas, I., Guan, H., Edwards, K., Nilsson, M., Varum, K.M., and Artursson, P. (2001) Chitosan as a nonviral gene delivery system. Structure–property relationships and characteristics compared with polyethylenimine in vitro and after lung administration in vivo. *Gene Therapy* **8**:1108–1121.

74. Hwang, S.J., Bellocq, N.C., and Davis, M.E. (2001) Effects of structure of beta-cyclodextrin-containing polymers on gene delivery. *Bioconjugate Chemistry* **12**:280–290.

75. Pun, S.H., Bellocq, N.C., Liu, A., Jensen, G., Machemer, T., Quijano, E., Schluep, T., Wen, S. et al. (2004) Cyclodextrin-modified polyethylenimine polymers for gene delivery. *Bioconjugate Chemistry* **15**:831–840.

76. Hedley, M.L. (2003) Formulations containing poly(lactide-co-glycolide) and plasmid DNA expression vectors. *Expert Opinion on Biological Therapy* 3:903–910.
77. Dhiman, N., Dutta, M., and Khuller, G.K. (2000) Poly (DL-lactide-co-glycolide) based delivery systems for vaccines and drugs. *Indian Journal of Experimental Biology* 38:746–752.
78. Singh, M., Fang, J.H., Kazzaz, J., Ugozzoli, M., Chesko, J., Malyala, P., Dhaliwal, R., Wei, R. et al. (2006) A modified process for preparing cationic polylactide-co-glycolide microparticles with adsorbed DNA. *International Journal of Pharmaceutics* 327:1–5.
79. Cheng, J., Teply, B.A., Sherifi, I., Sung, J., Luther, G., Gu, F.X., Levy-Nissenbaum, E., Radovic-Moreno, A.F. et al. (2007) Formulation of functionalized PLGA-PEG nanoparticles for in vivo targeted drug delivery. *Biomaterials* 28: 869–876.
80. Niola, F., Evangelisti, C., Campagnolo, L., Massalini, S., Bue, M.C., Mangiola, A., Masotti, A., Maira, G. et al. (2006) A plasmid-encoded VEGF siRNA reduces glioblastoma angiogenesis and its combination with interleukin-4 blocks tumor growth in a xenograft mouse model. *Cancer Biology and Therapy* 5:174–179.
81. Kushibiki, T., Nagata-Nakajima, N., Sugai, M., Shimizu, A., and Tabata, Y. (2005) Delivery of plasmid DNA expressing small interference RNA for TGF-beta type II receptor by cationized gelatin to prevent interstitial renal fibrosis. *Journal of Controlled Release* 105:318–331.

3

siRNA and DNA Transfer to Cultured Cells

Bagavathi Gopalakrishnan and Jon Wolff

Summary

Transfection is a powerful non-viral technology used to deliver foreign nucleic acids into eukaryotic cells, and is the method of choice for a variety of applications including studying the functional role of particular genes and the proteins they code for. By over-expressing genes to produce protein of interest and also by knocking down specific genes, researchers are able to accurately define the role of genes and the protein they encode in various cellular processes. Therefore, this powerful technology is a very vital component of the array of scientific research tools. However, the exact mechanism of action of transfection and also the numerous factors that influence the success of DNA or RNA delivery processes are not clearly understood. Hence, this chapter attempts to explain some of the popular cationic lipid/polymer-based transfection reagents for *in vitro* DNA/small inhibitory RNA (siRNA) delivery, mainly focusing on the protocols and critical factors to keep in mind to ensure successful delivery of nucleic acids into eukaryotic cells using these methods.

Key Words: Transfection; Non-viral DNA delivery; RNA delivery; siRNA delivery; Nucleic acid transfer; Knockdown; Reporter assay.

1. Introduction

Currently, the term transfection applies to the introduction of any naked nucleic acid molecule, not just DNA, into cultured eukaryotic cells, using non-viral means. This is a widely used, and often the most convenient technology to study the function of genes, to study the mechanism of gene regulation including interaction of factors that affect gene expression, to over-express genes, to produce protein of interest and thereby study protein function within eukaryotic cells, tissues, and organisms, and also to produce recombinant virus. When the

From: *Methods in Molecular Biology, vol. 480: Macromolecular Drug Delivery*, Edited by: M. Belting
DOI 10.1007/978-1-59745-429-2_3, © Humana Press, a part of Springer Science+Business Media, LLC 2009

foreign DNA is integrated into the chromosome then it is termed stable trans-fection, and this occurs with a relatively low frequency. The ability to select the positive clones, stable selection *(1)*, is made possible by using genes that encode resistance to an antibiotic, for e.g., gene neomycin phosphotransferase with the drug Geneticin. The cells that incorporate the foreign DNA into chromosomes can survive treatment with the drug and undergo clonal expansion that can be selected individually by 'ring-cloning' or by any other method of colony isola-tion, propagated, and characterized for protein expression *(2)*.

The development of reporter gene systems and also antibiotic-mediated selection methods for stable gene expression of the transfected DNA vastly expanded the applications for in vitro gene delivery technology. By using a non-endogenous reporter gene, for e.g., the bacterial chloramphenicol acetyltransferase (CAT) along with a sensitive detection system for the gene product, it allows the investigator to have the flexibility to clone upstream reg-ulatory elements to study how varying the conditions affects the expression of the reporter gene *(3)*. The reporter assay technology together with transfection reagents provides extremely versatile tools for studying in depth the promoter and enhancer sequences, *trans*-acting proteins such as transcription factors *(4)*, mRNA processing, protein–protein interactions, translation, and recombination events. Luciferase, β-galactosidase, alkaline phosphatase, and Green Fluores-cent Protein (GFP) reporter systems are other popularly used systems employed to study the above-mentioned processes.

Delivery of DNA, oligonucleotides, and RNA into mammalian cells is a critical step in transient gene expression studies, generation of stable cell lines, and pro-tein/virus production. Likewise, short interfering RNA (siRNA) delivery is vital for gene silencing studies using RNA interference (RNAi). RNAi technology is revolutionizing the biological research process as well as drug discovery and vali-dation. Using this technology one can turn gene expression 'off,' or knock it down, in order to better understand the function and role of a particular gene in disease or metabolic condition. The first step for this process is the successful transfection of siRNA. The success of transfections depends on choosing the right reagent to deliver with the highest efficiency leading to optimal production of protein from cells, or conversely inhibit expression of genes using siRNA, while avoiding cellu-lar toxicity. Effective transfection requires that a transfection reagent be efficient in delivering DNA or siRNA, yet be gentle to the cells.

It is important to maintain low cytotoxicity, since transfection reagent-mediated toxicity could potentially mask the true phenotype of a target gene being studied. This chapter will attempt to describe the use of several cationic lipid/polymer-based transfection reagents for the *in vitro* delivery of siRNA and DNA. Emphasis will be laid on the key parameters that affect transfection efficacy.

1.1. Transfection Technologies

In general, transfection methods can be divided into (*see* **Subheading 1.1.1.**) physical or direct transfer methods like electroporation, high-velocity bombardment, microinjection and (*see* **Subheading 1.1.2.**) chemical methods via carriers, e.g., lipofection, calcium phosphate, and DEAE-dextran.

1.1.1. Physical Methods

Physical methods include *Microinjection*, which is mainly used in single cell manipulations, such as oocytes injected with nucleic acids. This technique demands a certain level of skill to introduce a very thin micropipette into the cytoplasm or directly into the nucleus of an egg *(5)*. While it is not feasible to use this when dealing with multiple cells, the efficiency is nearly 100%. It is also a time-consuming and expensive method. *Biolistic particle delivery* is used frequently to introduce nucleic acids into in vitro as well as *in vivo* targets. The method involves coating of gold or tungsten microparticles with DNA or RNA, followed by acceleration most commonly by establishing a high-voltage discharge between two electrodes or gas pressure, followed by bombardment into cells *(6)*.

Electroporation is also a frequently used physical method for nucleic acid transfer. The cells and nucleic acids that are suspended in a special buffer are subjected to high-voltage pulses of electricity, which generates a potential difference across the membrane as well as charged membrane components and induces temporary pores in the cell membrane, enabling entry of nucleic acid *(7)*. It induces voltage-mediated rearrangement of membrane proteins resulting in hyper-porosity/leakiness of membranes and other functional changes, thereby allowing transfection of large DNA fragment. Although there is no reagent-induced toxicity and efficiencies are typically high, there is high-mortality rate, caused by high-voltage pulses or partial recovery of cells after the pulse. The critical phase in using this method is to ensure better survival rate of cells post-electroporation.

1.1.2. Chemical Methods

Chemical methods include a variety of popular DNA/RNA delivery techniques. DNA mixed with *DEAE-dextran* which is a polycationic derivative of dextran, a carbohydrate polymer that associates readily with the negatively charged nucleic acids, was one of the first reagents to be used for transient transfection. An excess of positive charge, contributed by the polymer allows the complex to come into closer association with the plasma membrane *(8)*. This method is simple and inexpensive, but is low in efficiency, in part due to high levels of toxicity. There are several other synthetic cationic polymers which have been used for DNA delivery into mammalian cells, including dendrimers, polybrene, and polyethyleneimine.

Calcium phosphate co-precipitation, first developed by Graham and van der Eb *(9)*, is popularly used for both stable and transient transfections in many different cell lines due to easy availability of component materials and their low cost. DNA when mixed with calcium chloride and added in a controlled manner to a buffered saline–phosphate solution at room temperature (RT) results in the formation of a precipitate that is dispersed on the attached cells *(10)*. The precipitate is presumably taken up by the cells by endocytosis. Calcium phosphate, like several other nucleic acid delivery reagents, protects the nucleic acid in the complex from the action of intracellular and serum nucleases.

Lipofection is the most popularly used chemical gene transfer method. One of the major advances in in vitro DNA delivery was the development of artificial liposomes for transfection. The next major advancement was the development of synthetic cationic lipids *(11)*. Lipofection transfection reagents can be divided into three different types: first generation comprises cationic liposomal reagents; the second generation consists of multicomponent liposomal reagents – lipids, polymers and combinations thereof. And the third generation consists of multicomponent liposomal reagents conjugated with antibodies or ligands which enable specific targeting *(12)*.

Cationic transfection lipids usually consist of a positively charged head group, such as an amine, a linker group that is characterized by flexibility, which is usually an ester or ether, and multiple hydrophobic tail groups *(13)*. A cationic lipid is mixed with a neutral lipid to form unilamellar liposome vesicles carrying a net positive charge to which nucleic acids get adsorbed. The positive charge facilitates an interaction with the plasma membrane, following which these complexes are endocytosed into the cell. Lipofection offers advantages such as relatively high efficiency, ability to transfect certain cell types that are regarded as difficult-to-transfect as well as the ability to transfer DNA as well as RNA and protein-to-animal and human cell lines. Also, both transient as well as stable lines can be created using DNA of various sizes.

Dendrimers are three-dimensional hyperbranched, globular macromolecules that are capable of condensing DNA into small complexes, and thereby increase plasmid transfection efficiency *(14)*. Dendrimers are typically stable in serum and not temperature sensitive, but are also non-biodegradable and cause significant cytotoxicity.

A large number of chemical methods to transfect nucleic acids have been developed with varying features and varying degrees of ease-of-use. However, the desirable features are common including high efficiency of transfer of foreign nucleic acid into the appropriate subcellular compartment of the cell to facilitate maximal by expression in the case of DNA and inhibition of gene expression in the case of siRNA.

1.2. RNAi

Introduction of exogenous siRNA into cells influences RNAi *(15)*. A huge percentage of transcriptional output of the genome includes non-protein coding RNA. RNAi technology takes advantage of the cells endogenous machinery, facilitated by short interfering RNA molecules, to effectively knockdown expression of a gene of interest. Among the several ways to induce RNAi are synthetic RNA molecules, RNAi vectors, and in vitro dicing. siRNA are short pieces of double-stranded RNA (dsRNA) in the cell that, initiate the specific degradation of a targeted cellular mRNA. In this process, the antisense strand of the siRNA becomes part of a multiprotein complex, or RNA-induced silencing complex (RISC), which then identifies the corresponding mRNA and cleaves it at a specific site. The cleaved message is then targeted for degradation, which correlates to a reduction in gene expression.

Table 1
List of Some of the Cationic Lipid/Polymer-Based siRNA Transfection Reagents Currently Available

Transfection Reagent	Manufacturer
Lipofectamine RNAiMAX	Invitrogen Corporation
siLentFect	Bio-Rad
siPORT Lipid	Ambion, Inc.
siPORT Amine	Ambion, Inc.
RNAiFect	Qiagen, Inc.
TransMessenger	Qiagen, Inc.
DharmaFECT 1,2,3, and 4	Dharmacon
X-tremeGENE siRNA	Roche-Applied Science
TransPass R2	New England Biolabs
siIMPORTER	Upstate Cell Signaling Solutions
TransIT-TKO	Mirus Bio Corporation
TransIT-siQUEST	Mirus Bio Corporation
GeneSilencer	Gene Therapy Systems, Inc.
siFECTamine	IC-VEC
jetSI	PolyPlus-Transfection SAS
petSI-ENDO	PolyPlus-Transfection SAS
GeneEraser	Stratagene

MicroRNAs (miRNAs) are non-coding, single-stranded RNA (ssRNA) molecules of about 21–23 nucleotides and are thought to down-regulate the expression of other genes *(16)*. Both RNAi and short hairpin RNA (shRNA) systems take advantage of the endogenous RNAi pathway found in eukaryotic cells. Compared to shRNA vectors, RNAi vector systems, using artificial miRNAs, utilize more of the components of the endogenous cellular machinery, resulting in more efficient processing of expressed RNA hairpins *(17)*. Information on designing siRNA is not dealt within this chapter and can be found elsewhere *(18)*.

RNAi technology can be used to assess the function of many genes within the genome that are potentially involved in diseases. It provides an effective means of blocking expression of a specific gene and evaluating its response to chemical compounds or changes in signaling pathways. Especially in comparison to other nucleotide-based methods, such as antisense or ribozymes, RNAi is a specific, potent, and highly successful approach for loss-of-function studies. However, to accomplish efficient target gene knockdown, siRNA must be delivered to a high percentage of the cells in the population. Therefore, choosing the most appropriate transfection reagent and optimizing the protocol become critical to the success of the experiment. A number of reagents are currently available to transfect siRNA and are listed in **Table 1**.

2. Materials

2.1. Cell Culture

1. Complete culture medium appropriate for the cell line being used, including Dulbecco's Modified Eagle's Medium (DMEM) or Modified Eagle's Medium (MEM) or any such growth medium supplemented with 10% fetal bovine serum (FBS, Hyclone, Ogden, UT) and antibiotics (Penicillin–Streptomycin, Invitrogen, CA)
2. Cell lines (e.g., HepG2, HeLa-ATCC, VA)
3. Phosphate Buffered Saline (PBS, Hyclone, Ogden, UT)
4. Trypsin (0.25%) with 1 mM ethylenediamine tetraacetic acid (EDTA) (Invitrogen, CA)
5. Tissue culture plates (Nunc – NY, Becton & Dickinson – NJ, Corning – NY)
6. Spinner Flasks (125 mL) (Corning, NY), multistirrer (Thermolyne, IA)
7. Sterilgard III Advance tissue culture laminar flow hood (Baker Co., ME)
8. Napco humidified CO_2 incubator (Thermo Electron, MA)
9. Inverted microscope (Zeiss, NY)

2.2. Transfection

1. Transfection reagent such as TransIT®-LT1 (Mirus Bio, Madison, WI)
2. Polystyrene plastic tubes, pipette-tips (Becton & Dickinson, NJ)
3. Pipetman (Rainin instruments, CA)
4. OptiMEM® (Invitrogen, CA)

5. Plasmid DNA expression vector such as pLIVE®-lacZ (Mirus Bio, WI), or pCI-luc (Promega, WI)

2.3. Reporter Assay

1. Lysis buffer and luminol substrate (Promega, WI)
2. Beta-Gal staining kit (Mirus Bio, WI)
3. Veritas Microplate Luminometer (Promega, WI)

2.4. siRNA Transfection

1. siRNA Transfection reagent such as TransIT-TKO® (Mirus Bio, WI)
2. Custom synthesized siRNA specifically targeted against the gene to be knocked down (Dharmacon, CO)

3. Methods
3.1. DNA Transfection of Adherent Cells

Due to the vast number of DNA transfection methods and reagents available to the researcher (**Table 2**), it will be a hard task to describe specific protocols

Table 2
List of Some of the Cationic Lipid/Polymer-Based DNA Transfection Reagents Currently Available

Transfection Reagent	Manufacturer
Lipofectamine 2000	Invitrogen Corporation
Lipofectamine-LTX	Invitrogen Corporation
FreeStyle MAX Reagent	Invitrogen Corporation
Fugene 6	Roche-Applied Science
Fugene HD	Roche-Applied Science
TransFectin	Bio-Rad
ProteoJuice	Novagen, Inc.
GenePorter	Genlantis
HiFect	Amaxa, Inc.
Effectene	Qiagen, Inc.
SuperFect	Qiagen, Inc.
Jet-PEI	Polyplus
GeneJammer	Stratagene
Turbofectin 8.0	Origene
PrimFect II	Cambrex BioScience
TransIT-LT1	Mirus Bio Corporation
TransIT-293	Mirus Bio Corporation
TransIT-Express	Mirus Bio Corporation

recommended for each reagent. However, many protocols share many common themes. For demonstration purposes, this chapter shall describe in detail the use of one DNA delivery reagent, namely, the TransIT-LT1 transfection reagent (**Table 2**), a broad spectrum DNA transfection reagent, which efficiently delivers DNA into many cell types.

3.1.1. Cell Plating

The day prior to transfection, plate healthy, actively dividing cells at a cell density of $1–3 \times 10^5$ cells in complete growth medium per well of a 6-well plate to obtain 50–80% confluence the following day. The actual number of cells to be plated is determined empirically, since it varies depending on growth characteristics and morphology of the adhered cell. Besides the cell line, several transfection reagents work best under different cell densities than the range mentioned, hence it is important to adhere to manufacturer's recommended conditions. Incubate at the appropriate cell culture conditions, in a humidified tissue culture incubator at 37°C and 5% CO_2.

3.1.2. Complex Formation

This procedure is performed immediately prior to transfection.

The cells are examined on the day of transfection to make sure that they are at the optimal cell density (60–80%). Add the TransIT®-LT1 reagent (2–8 μL per 1 μg DNA) directly into 250 μL of serum-free medium contained in a sterile polystyrene tube. Mix completely by gentle pipetting. Incubate the reagent at RT for 10 min and add plasmid DNA (pCI-luc or pLIVE-lacZ) (1–3 μg per well) to the diluted TransIT®-LT1 reagent and mix by gentle pipetting, followed by incubation at RT for 15–30 min (*see* **Note 3**).

3.1.3. Harvest and Reporter Assay

If necessary, remove the medium from the cells prepared in **Subheading 3.1.1.** and replace them with 2 mL of fresh complete growth medium per well of a 6-well plate. Add the transfection reagent/DNA complex mixture, prepared in **Subheading 3.1.2.**, dropwise to the cells in complete growth medium. Gently rock the dish back and forth and from side-to-side to distribute the complexes evenly. Incubate for 24–48 h. The above incubation is designed for transfections performed with no media change. To perform a media change after a 4–24 h-incubation with the complexes, replace the original medium with fresh complete growth medium, and incubate for an additional 24–48 h. Harvest cells and assay for reporter gene activity. Lyse cells using Promega lysis reagent and measure relative luciferase units if transfecting pCI-luc or use β-galactosidase staining kit to measure pLIVE®-lacZ expression.

3.2. Transient Transfection of Suspension Cells in 6-Well Plates

This method is used to transiently transfect cells in suspension like Jurkat cells.

3.2.1. Cell Plating

Approximately 24 h prior to transfection, plate cells at a cell density of $\sim 10^6$ cells in complete growth medium per well of a 6-well plate and culture the cells overnight.

3.2.2. Complex Formation (Perform This Procedure Immediately Prior to Transfection)

Add the TransIT®-LT1 reagent (2–8 μL per 1 μg DNA) directly into 250 μL of serum-free medium contained in a sterile polystyrene tube. Mix by gentle pipetting. For transfecting larger amounts of DNA, or if a precipitate forms upon adding the DNA, increase the volume of serum-free medium up to 1 mL, followed by incubation at RT 10 min. Add plasmid DNA (1–3 μg per well) to the diluted TransIT®-LT1 reagent and mix completely by gentle pipetting and incubate at RT for 20–30 min.

3.2.3. Harvest and Reporter Assay

Spin down cells and remove media. Lyse cells and perform reporter assay as in **Subheading 3.1.3.** to assay for relative light units as readout for luciferase activity.

3.3. Transient Transfection of Suspension Cells in Spinner Flasks

The following protocol describes the method to transfect HEK-293 and CHO cells fully adapted to suspension in serum-free media.

3.3.1. Cell Seeding

Approximately 18–20 h prior to transfection, seed cells at a cell density of 5×10^5 cells per mL in serum-free growth medium. Seed 25×10^6 cells in 50 mL of culture media in a 125 mL Corning Spinner Flask to reach approximately double the density overnight.

3.3.2. Cell Viability

Use 0.4% Trypan Blue to stain cells and count viable cells at seeding. Viability should be at least 95% at this stage. Culture cells in 50 mL media in Corning Spinner Flasks at 8% CO_2, 37°C, in a humidified incubator (*see* **Note 1**).

3.3.3. Complex Formation

Immediately prior to transfection, the transfection complex is prepared. Add the TransIT®-LT1 reagent (6 μL per 2 μg DNA) directly into 200 μL OptiMEM contained in a sterile polystyrene tube per mL of culture medium. Mix by gentle pipetting. For transfection of approximately 50×10^6 cells in 50 mL culture media dilute 300 μL of TransIT®-LT1 in 10 mL OptiMEM and incubate at RT for 10 min. (See Note 3) Add plasmid DNA (2 μg per mL of culture medium) to the diluted TransIT®-LT1 reagent and mix completely by pipetting, followed by incubation at RT for 20 min.

3.3.4. Transfection

Add the TransIT®-LT1 reagent/DNA complex mixture prepared in **Subheading 3.3.3.** to the cells and place flasks immediately on slow speed stir-plate set at 70–80 rpm inside an 8% CO_2, 37°C humidified incubator.

3.3.5. Cell Harvest and Assay

Incubate cells with transfection complex for 24 h. Test cell viability using 0.4% Trypan Blue to stain cells and count viable cells at seeding, day of transfection, and just prior to harvest. Harvest cells and assay for reporter gene activity: Spin down cells and lyse cells to assay for luciferase activity. **Figures 1 and 2** illustrate typical transfection results, observed visually as lacZ stained cells (**Fig. 1**) or Relative Light Units (RLUs) when luciferase activity is measured (**Fig. 2**).

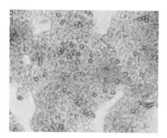 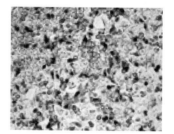

A. Control-Untransfected cells.   B. Cells transfected with plasmid DNA.

Fig. 1. Transfection of plasmid DNA with TransIT-LT1 into HepG2 cells: HepG2 cells were grown to 80% confluency in 12-well tissue culture plates and transfected with 1 μg of pLIVE-lacZ using 3 μL TransIT-LT1 per well. Twenty-four hours post-transfection the cells were fixed and stained for β-galactosidase using Beta-Gal staining kit. The transfected cells (B) appear dark in color, while the control-untransfected cells (A) do not stain. Images taken at 60 × magnification.

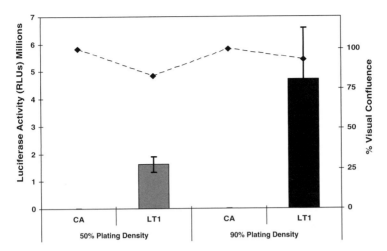

Fig. 2. TransIT-LT1-mediated transfection of HeLa cells at varying cell densities: HeLa cells were plated in 12-well plates and transfected in parallel at 50% and 90% confluency. TransIT-LT1 reagent transfections were performed in duplicate using a luciferase expression vector (pCI-luc) and 3 µL reagent per well, shows the importance of plating cells at optimal density for transfection. Twenty-four hours post-transfection, cells were harvested and assayed for luciferase activity. Visual confluence (*line graph*) was measured under the microscope at harvest. The data represent the average luciferase activity (Relative Light Units – RLUs in Millions) from three experiments performed on different days.

3.4. RNAi Transfection to Knockdown Expression of Endogenous Target Gene

The discovery of RNAi has enabled researchers to use siRNA to rapidly and specifically knockdown gene expression in a variety of mammalian cells *(19)*. Most of the methods and techniques to deliver DNA have been used successfully to deliver siRNA into mammalian cells, but there are transfection reagents that have been optimized for siRNA delivery *(20)*. As with transfection reagents there are multiple lipid-based reagents (**Table 2**) at the disposal of the researcher to carryout RNAi experiments. For the purpose of demonstrating the technique, we have used TransIT-TKO® (Mirus Bio, Madison, WI) to illustrate the salient features of RNAi experiments and methodology to be adopted while performing gene knockdown experiments.

3.4.1. Cell Plating of Adherent Cells

Approximately 24 h prior to transfection, plate cells at an appropriate cell density to obtain ~60–80% confluence the following day. Plate 6×10^4–2.4×10^5

cells per mL adherent cells in 500 μL of complete growth media per well (*see* **Note 1**). Incubate the cells overnight.

3.4.2. Cell Plating of Suspension Cells

Immediately prior to transfection, plate cells in 250 μL of complete growth media per well of a 24-well plate, at an appropriate density of 6×10^5–1×10^6 cells per mL. Alternatively, plate 250 μL of cells the day prior to transfection at an appropriate density of 3–5×10^5 cells per mL.

3.4.3. Complex Formation

Perform this procedure immediately prior to transfection. Add 50 μL of serum-free media such as OptiMEM I contained in a sterile plastic tube. Add 1–4 μL of TransIT-TKO reagent directly into the serum-free media. Mix thoroughly by pipetting. Add 10–50 nM siRNA (the starting recommendation is 25 nM final concentration in the well) to the diluted transfection reagent. Mix by gentle pipetting and incubate at RT for 5–20 min (*see* **Note 12**).

3.4.4. Cell Preparation for Transfections in Complete Growth Media

For adherent cell types, adjust the volume in the well to 250 μL of complete growth media by removing ~250 μL (half) of the original plating media to conserve siRNA. This media reduction requires the use of less siRNA to obtain the recommended 25 nM (final concentration in the well). Add the transfection reagent/siRNA complex mixture prepared (*see* **Subheading 3.4.3.**), dropwise to the cells. Gently rock the plate back and forth and from side-to-side to distribute the complexes evenly. Do not swirl the plate and incubate for 24–72 h (*see* **Note 4**).

3.4.5. Assay for Knockdown of Target Gene Expression

Assay using relative light units produced by luciferase as the read out by the method described for DNA transfection reporter assay in **Subheading 3.3.5.** Other methods of choice to determine gene knockdown are Northern blots, RT-PCR, RNase protection, and branched DNA assays, apart from which assays to measure target protein produced can also be used.

3.5. Sequential Transfection of Plasmid DNA in a T-75 Flask and siRNA in 24-Well Plates (for Adherent Cells)

Simultaneous transfections using the TransIT®-LT1 reagent (for DNA delivery) and TransIT-TKO® reagent (for siRNA delivery) can be performed (*see* **Note 7**). The following recommendations are for plasmid DNA transfection of adherent cells in a T-75 flask. The plasmid transfected cells are then re-plated in a 24-well plate and subsequently transfected with siRNA.

3.5.1. Cell Plating of Adherent Cells (in a T-75 Flask)

Approximately 24 h prior to transfection, plate 6×10^4–2.4×10^5 cells per mL adherent cells in a T-75 flask, to attain an appropriate cell density of ~60–80% confluence the following day. Incubate the cells overnight.

3.5.2. Complex Formation: TransIT®-LT1 Reagent and Plasmid DNA (pDNA)

The following protocol recommends the TransIT®-LT1 transfection reagent. In a sterile polystyrene tube, add 500 μL of serum-free medium. Add 45 μL of TransIT®-LT1 transfection reagent directly into the serum-free medium. Mix thoroughly by pipetting. Add 15 μg plasmid DNA (titration of pDNA may be necessary) to the diluted TransIT®-LT1 reagent. Mix by gentle pipetting (*see* **Note 3**), and incubate at RT for 15–30 min.

3.5.3. Plasmid Transfection

Add the TransIT®-LT1 reagent/DNA complex mixture (from **Subheading 3.5.2.**) dropwise to the cells in complete growth medium. Gently rock the flask back and forth and from side-to-side to distribute the complexes evenly. Do not swirl the flask and incubate for 2–4 h.

3.5.4. Transfer of Cells from T-75 Flask(s) to a 24-Well Plate(s)

Trypsinize the T-75 flask according to standard procedures (*see* **Note 9**) and add 14 mL of complete growth media to the T-75 flask. Mix cells thoroughly. Plate cells (500 μL per well of a 24-well plate) at an appropriate cell density (1.2×10^4–4.8×10^5 cells per mL) to obtain ~60–80% confluence at transfection and incubate for 2–3 h to allow cells to adhere to the plate.

3.5.5. Complex Formation: Transfection Reagent and siRNA

The complex consisting of the transfection reagent and siRNA is formed by the same procedure as in **Subheading 3.5.2.**

3.5.6. siRNA Transfection

Add the transfection complex to the cells, following which assay of knock-down is performed as previously described in **Subheading 3.4.5.**

3.6. Sequential Transfection of Plasmid DNA and siRNA in 24-Well Plates (for Suspension Cells)

3.6.1. Cell Plating of Suspension Cells

Immediately prior to transfection, plate cells in 250 μL of complete growth media per well of a 24-well plate, at an appropriate density (such as 6×10^5– 1×10^6 cells per mL). Alternatively, plate 250 μL of cells the day prior to transfection at an appropriate density (such as 3–5×10^5cells per mL).

3.6.2. Complex Formation: TransIT®-LT1 Reagent and Plasmid DNA (pDNA)

Add 50 μL of serum-free medium to a sterile polystyrene tube. Add TransIT®-LT1 transfection reagent (1–2 μL per well for 24-well plates; scale up or down for different size plates) directly into the serum-free medium. Mix thoroughly by pipetting or vortexing. Add plasmid DNA (0.5 μg per well for 24-well plates; titration of pDNA may be necessary) to the diluted TransIT®-LT1 reagent. Mix by gentle pipetting and incubate at RT for 15–30 min.

3.6.3. Plasmid Transfection in Suspension Cells

Add the TransIT®-LT1 reagent/DNA complex mixture dropwise to the cells in complete growth medium. Gently rock the plate back and forth and from side-to-side to distribute the complexes evenly and incubate for 2–4 h.

3.6.4. Complex Formation: TransIT-TKO® Reagent with siRNA

Pipette 50 μL of serum-free media into a sterile polystyrene tube. Add 0.5–4 μL of TransIT-TKO® reagent directly into the serum-free media. Mix thoroughly by pipetting. Add 10–50 nM siRNA to the diluted transfection reagent. Mix by gentle pipetting and incubate at RT for 5–20 min.

3.6.5. siRNA Transfection in Suspension Cells

Add the transfection reagent/siRNA complex mixture dropwise to the cells. Gently rock the plate back and forth and from side-to-side to distribute the complexes evenly and incubate for 24–72 h. Finally, perform assay for inhibition of target gene expression as described in **Subheading 3.4.5.**

3.7. Simultaneous Transfection of DNA and siRNA Using TransIT-TKO Reagent in 24-Well Plates

This protocol elaborates simultaneous transfections using the TransIT-TKO® reagent (for siRNA delivery) and TransIT®-LT1 reagent (for DNA delivery)

3.7.1. Cell Plating of Adherent Cells

The day prior to transfection, plate 500 μL of cells at an appropriate cell density to obtain ~60–80% confluence the following day (6×10^4–2.4×10^5 cells per mL, depending on cell size and characteristics) and incubate the cells overnight.

3.7.2. Cell Plating of Suspension Cells

Immediately prior to transfection, plate cells in 250 μL of complete growth medium per well of a 24-well plate, at an appropriate density (such as 6×10^5–1×10^6 cells per mL). Alternatively, plate 250 μL of cells the day prior to transfection at an appropriate density (3–5×10^5 cells per mL).

3.7.3. Complex Formation

Add 50 μL of serum-free medium into a sterile polystyrene tube. Add TransIT®-LT1 transfection reagent (1–2 μL per well for 24-well plates; scale up or down for different size plates). Add plasmid DNA (0.5 μg per well for 24-well plates; titration may be necessary) to the diluted transfection reagent. Mix by gentle pipetting (*see* **Note 8**). Incubate at RT for 15–30 min. In the same tube, add the TransIT-TKO® reagent (1–4 μL per well for 24-well plates) and mix by gentle pipetting. Add siRNA (10–50 nM) to the diluted complex mixture and mix by gentle pipetting. Incubate at RT for 5–20 min.

3.7.4. Dual Transfection

For adherent cell types, adjust the volume in the well to 250 μL of complete growth medium by removing ~250 μL (half) of the original plating medium to conserve siRNA. This media reduction requires the use of less siRNA to obtain the recommended 25 nM (final concentration in the well). Add TransIT®-LT1 reagent/pDNA/TransIT-TKO® reagent/siRNA complex mixture dropwise to the cells and incubate for 24–72 h. Finally, perform assay for inhibition of target gene expression as described in **Subheading 3.4.5**.

4. Conclusion

As is evident by the numerous choices available to the user, there are many ways to achieve the goal of transfection of DNA or siRNA into a mammalian cell. Depending on the cell line and other experimental conditions, especially keeping in view the downstream application of the transfection procedure, the researcher has to choose the optimal reagent and the appropriate methodology to achieve maximum efficiency.

Some of the challenges in performing efficient transfection are as described in detail in the chapter, to reduce toxicity, achieve higher efficiency without deleterious manipulation of gene expression machinery, reduce lipid-induced off target effects, and interferon responses. The procedure of transfection is still relatively poorly understood and the challenges posed to the researcher are still daunting at times. The objective of this chapter is to equip the user with vital information to enhance the success of transfection experiments.

5. Notes

1. Most protocols recommend cell confluences between 50% and 80% at the time of transfection. Some protocols such as Lipofectamine 2000 recommend higher confluences. It is important to be in the optimal range for the respective protocols. As the optimal cell density for efficient transfection can vary between cell types, maintain the same seeding protocol between experiments. Use conditions appropriate for the cell type being transfected. Standard incubation conditions for mammalian cells are 37°C in 5% CO_2. Other cell types, such as insect cells, require different temperatures and different CO_2 concentrations. Proper cell handling during harvesting process is imperative to maintain the integrity of the cell and ensure the success of the experiment. Before harvesting, it is better to wash the cells off of any residual growth medium, as well as, calcium and magnesium ions using PBS or Hank's buffered salt solution (HBSS) without these ions. Do not scrape the monolayer unless specifically recommended by the cell supplier. Scraping of the cell layer may cause mechanical damage to the cells and will not result in a single cell suspension.

2. Some of the critical factors to be considered while choosing the transfection reagent and while designing the experiment are target molecule to be delivered, length of expression, cell type, cellular context, desired viability, desired efficiency, expertise, cost, and time.

3. Many transfection reagents yield improved transfection efficiencies when transfections are performed in complete growth medium (instead of serum-free medium) and the media change is eliminated. Whereas in some cases a media change is necessitated because of toxicity. Hence, it is critical to change media while using these reagents.

    OptiMEM I reduced-serum medium is widely used for complex formation. Other serum-free media can also be used as long as they are free of polyanions such as heparin or dextran-sulfate, which compete with DNA or nucleic acids to bind to the transfection reagent.

4. Incubation time of siRNA with reagent determines the particle size of the transfection complex, which varies depending on cell lines. Therefore, it is an important variable to be determined for each particular cell line.

5. The indicated incubation time is designed for transfections performed with no media change. If a media change is necessary to remove the transfection

complexes, incubate the cells for 24 h, replace the original media with fresh complete growth media, and incubate for an additional 24–48 h.

6. Suspension cells should be transfected when they are in the logarithmic growth phase which is generally, 10^6–10^7 cells per mL. It is important to avoid extra pipetting or unnecessary washing steps. Do not vortex cells.

7. Some reagents like TransIT-TKO® can be used for simultaneous transfection of DNA and siRNA, since this reagent does not inhibit DNA delivery. Others like the TransIT®-siQUEST reagent cannot be used in tandem. Hence, the user has to be aware of the mutual compatibility of the reagents while designing the knockdown experiment.

8. If transfecting more than one plasmid, mix the plasmids together in a microcentrifuge tube and incubate for 5 min at RT before adding to the diluted transfection reagent.

9. Transfection reagents can cause cells to adhere, therefore, ensure cells are completely removed from the bottom of the flask by trypsinisation.

10. RNAi vectors use Pol II to express artificial miRNAs that are designed to cleave target genes. short hairpin RNAi (shRNAi) Vectors use Pol III promoters to express short hairpin RNAs that cleave target of interest.

11. Using a negative control consisting of the siRNA Delivery Reagent, and a non-specific siRNA such as Dharmacon's siCONTROL Non-Targeting siRNA #1 in serum-free medium is recommended. Other controls to consider are cells alone control and a reagent alone control. Perform these controls in parallel to the serum-free medium/transfection reagent/specific siRNA transfections using the same volumes and transfection parameters.

12. Apart from standard optimization as performed for DNA transfections additional considerations for siRNA transfections include optimizing siRNA concentration – siRNA used for transfection should be pure, sterile, and have the correct sequence. As a starting point, use 25 nM siRNA (final concentration in the well). Depending on the type of experiment, the optimal final siRNA concentration for transfection is typically within the range of 10–50 nM.

13. *Troubleshooting for low transfection efficiency:*

    (a) **Suboptimal transfection reagent to DNA ratio:** Determine the optimal transfection reagent to DNA ratio by titrating the reagent from 2 to 8 μL per 1 μg DNA. Choose the amount which gives the highest transfection efficiency and the lowest cellular toxicity.

    (b) **Cell density (% confluence) not optimal at the time of transfection:** The recommended cell density for most cell types at the time of transfection is 50–70% confluence. However, it may be necessary to determine the optimal cell density for each cell type in order to maximize transfection efficiency. Maintain this density in future experiments to ensure reproducibility.

    (c) **Poor quality of transfecting DNA (DNA may be partially degraded, or an inhibitor, such as endotoxin, may be present in the preparation):** DNA used for transfection should be highly purified, sterile, and free from contaminants such as endotoxin. Endotoxins are negatively charged and associate with positively charged proteins that copurify with DNA. These are toxic to cells as they cause septic shock. The bacterial lipopolysaccharides which constitute the endotoxins are removed by solubilizing them with detergents, as done in most endotoxin removal kits. The purified DNA should be resuspended in sterile deionized water or TE (Tris–EDTA) buffer before use. Using phenol:chloroform during preparation of DNA can also be harmful, as these are toxic to living cells and very difficult to remove completely.

(d) **Fetal calf serum present during transfection reagent/DNA complex formation:** Use serum-free medium when forming the complexes. Presence of serum affects complex formation.

(e) **Inhibitor present during transfection:** The presence of polyanions, such as dextran sulfate or heparin, can inhibit transfection. Use transfection medium that does not contain these polyanions.

(f) **Cell morphology has changed:** If the cell passage number is too high or too low the transfection efficiency may be adversely affected. Maintain a similar passage number between experiments to ensure reproducibility.

14. *Troubleshooting for high cellular toxicity:*

(a) **Complexes were added to the cells in serum-free media:** Form complexes in serum-free media, and add to cells in complete growth media (serum containing). Transfection efficiency is improved and cytotoxicity is decreased when the complexes are added to cells in complete growth media and the media change is eliminated.

(b) **Cell density (% confluence) was not optimal at the time of transfection:** Addition of transfection complex to cells below the recommended cell density could lead to cytotoxicity. Therefore, it is necessary to determine the optimal cell density for each cell type in order to reduce toxicity.

(c) **Excessive amount of transfection reagent/DNA complex mixture was in the transfection:** Reduce the amount of transfection reagent or DNA added to the cells.

(d) **Reagent/DNA complex was not mixed thoroughly with the cells in the well plate:** Mix thoroughly to evenly distribute the complexes to all cells. Rocking the dish back and forth and from side-to-side is recommended. Do not swirl or rotate the dish, as this may result in uneven distribution.

(e) **Using cells which have too low or too high passage number:** If the passage number of the cells is too high or too low, they can be more sensitive to transfection reagents causing excessive cell death.

15. *Optimization of transfection conditions:*

The key to successful transfection is careful optimization of reaction conditions for each individual cell type. Cells of a lower passage number typically respond better to transfection and ensure higher efficiency than those with higher passage number. Also, some cell lines differentiate and change their features after many passages. Conversely, it is best to allow cells to settle down into a comfortable growth pattern soon after they are thawed from cryo-preserved state. A couple of passages of culturing ensure that. Cells should also be treated carefully and a regular pattern of growing and splitting and a middle range of cell density – neither too high nor low, ensures that cells are relatively easier to transfect. To ensure efficient transfections certain variables affecting transfection have to be controlled by a logical optimization procedure.

(a) **Media conditions:** Certain reagents yield improved transfection efficiencies when transfections are performed in complete growth medium (instead of serum-free medium) and the media change is eliminated.

(b) **Cell density (% confluence) at transfection:** The recommended cell density for most cell types at transfection is 50–90% confluence. Determine the optimal cell density for each cell type in order to maximize transfection efficiency. Maintain this density in future experiments for reproducibility.

(c) **Passage number and morphology of cells:** The state of the cells at the time of transfection could have a dramatic impact on efficiency. The passage number and cellular morphology have to be monitored closely and vary from cell line-to-cell line, but in general it is better to use cells younger than 50 passages for transfection. And if cells start to transfect poorly it is advised to discard those cells and begin with freshly cultured, lower passage number cells from frozen stock.

(d) **DNA purity and concentration for transfection:** DNA used for transfection should be highly purified, sterile, and most importantly free from contaminants such as endotoxin. Remove any traces of endotoxin (bacterial lipopolysaccharide) using endotoxin removal kit. The optimal DNA concentration for transfection is 1–3 μg per well of a 6-well plate. As a starting point, 2.5 μg per well of a 6-well plate is recommended.

(e) **Transfection reagent to DNA ratio:** As a starting point, use 3 μL of TransIT®-LT1 reagent per 1 μg of DNA. Titrate the TransIT®-LT1 reagent from 2 to 8 μL per 1 μg DNA, depending on the specific cell type. For future transfections, use the ratio that gives the best transfection efficiency with the lowest cellular toxicity, on similarly passaged cells.

(f) **Transfection incubation time:** Determine the optimal incubation time empirically by testing a range from 4 to 48 h.

(g) **Transfection of hard-to-transfect and primary cells:** Some cells are traditionally less permissible to transfection, hence termed 'hard-to-transfect.' And to overcome the resistance of these cells, the dose needed is generally in the higher order of magnitude. Some lymphoid cell lines and primary cells, including human umbilical vascular cells (HUVEC), RAW264.7 macrophages, neural progenitor cells, and human mesenchymal stem cells, fall under this category of cells. Electroporation in most cases is the most efficient method of gene delivery into hard-to-transfect cell-lines such as those listed above.

16. *Troubleshooting for siRNA transfection-mediated knockdown:*
Low knockdown efficiency

(a) **Suboptimal volume of transfection reagent:** Use the optimal volume of reagent that gives the highest knockdown efficiency with the lowest cellular toxicity.

(b) **Suboptimal siRNA concentration:** Determine the optimal siRNA concentration by titrating from 10 to 50 nM, starting with a final concentration in the well of 25 nM of siRNA. In some instances, higher concentrations of siRNA such as 200 nM may be necessary to achieve sufficient knockdown of the gene of interest, but generally delivery of high levels of siRNA (>100 nM) has been shown to alter cellular gene expression profiles non-specifically *(21)*. Therefore, use the lowest possible amount of siRNA necessary to produce the desired knockdown, avoiding potential non-specific effects.

(c) **Low transfection efficiency:** Follow transfection protocol steps carefully as described earlier in the chapter.

(d) **Denatured siRNA:** To dilute siRNA, use the manufacturer's recommended buffer, or 100 mM NaCl, 50 mM Tris, pH 7.5 in RNase-free water. Do not use water as this can denature the siRNA duplex.

(e) **Incorrect siRNA sequence:** Ensure that the sequence of siRNA is correct for the gene of interest. More than one sequence may need to be tested for optimal knockdown efficiency.

(f) **Poor quality of transfecting siRNA:** Avoid siRNA degradation by using RNase-free handling procedures and plasticware. Degradation of siRNA can be detected on acrylamide gels.

(g) **Fetal calf serum present during complex formation:** Use serum-free medium during complex formation steps.

(h) **Cell density (% confluence) not optimal at the time of siRNA transfection:** The recommended cell density for most cell types at the time of siRNA transfection is 60–80% confluence. Determine the optimal cell density, which may be outside the recommended range, for each cell type in order to maximize transfection efficiency. Maintain the optimal density in future experiments for reproducibility.

(i) **Proper controls were not included:** To ensure assay validation include the following controls: cells only (for visual comparisons), transfection reagent alone, and transfection reagent with a non-specific siRNA, such as Dharmacon's siCONTROL Non-Targeting siRNA #1. To verify efficient transfection and knockdown, deliver an siRNA targeted against an endogenous gene, such as GAPDH or Lamin A/C, followed by western blotting or target mRNA quantification.

References

1. Ausubel, F.M., Brent, R., Kingston, R.E., Moore, D.D., Seidman, J.G., Smith, J.A., and Struhl, K. (1991) *Current Protocols in Molecular Biology*, (2nd edn.), Vol. 1. John Wiley and Sons, New York.

2. Zeyda, M., Borth, N., Kunert, R., and Katinger, H. (1999) Optimization of sorting conditions for the selection of stable, high-producing mammalian cell lines. *Biotechnol Prog.* **15**:953–957.
3. Agarwal, A., Halvorson, L.M., and Legradi, G. (2005) Pituitary adenylate cyclase-activating polypeptide (PACAP) mimics neuroendocrine and behavioral manifestations of stress: Evidence for PKA-mediated expression of the corticotropin-releasing hormone (CRH) gene. *Brain Res Mol Brain Res.* **29**; **138**:45–57.
4. Dai, H., Hogan, C., Gopalakrishnan, B., Torres-Vazquez, J., Nguyen. M., Park, S., Raftery, L.A., Warrior, R., and Arora, K. (2000) The zinc finger protein schnurri acts as a Smad partner in mediating the transcriptional response to decapentaplegic. *Dev Biol.* **227**:373–387.
5. Recillas-Targa, F. (2006) Multiple strategies for gene transfer, expression, knockdown, and chromatin influence in mammalian cell lines and transgenic animals. *Mol Biotechnol.* **34**:337–354.
6. Dileo, J., Miller, T.E., Chesnoy, S., and Huang, L. (2003) Gene transfer to subdermal tissues via a new gene gun design. *Hum Gene Ther.* **14**:79–87.
7. Xie, T., and Tsong, T.Y. (1993) Study of mechanisms of electric field-induced DNA transfection. V. Effects of DNA topology on surface binding, cell uptake, expression and integration into host chromosomes of DNA in the mammalian cell. *Biophys J.* **65**: 1684–1689.
8. Pari, G.S., and Xu, Y. (2004) Gene transfer into mammalian cells using calcium phosphate and DEAE-dextran. *Methods Mol Biol.* **245**:25–32.
9. Graham, F.L., and Van der Eb, A.J. (1973) A new technique for the assay of infectivity of human adenovirus 5 DNA. *Virology* **52**:456–467.
10. Sambrook, J., Fritsch, E.F., and Maniatis, T. (1989) *Molecular Cloning: A Laboratory Manual*, (2nd edn.). Cold Spring Harbor Laboratory, Cold Spring Harbor, New York.
11. Felgner, P.L., Gadek, T.R., Holm, M., Roman, R., Chan, H.W., Wenz, M., Northrop, J.P., Ringold, G.M., and Danielsen, M. (1987) Lipofection: a highly efficient, lipid-mediated DNA-transfection procedure. *Proc Natl Acad Sci USA.* **84**:7413–7417.
12. Wei, Y., Pirollo, K.F., Yu, B., Rait, A., Xiang, L., Huang, W., Zhou, Q., Ertem, G., and Chang, E.H. (2004) Enhanced transfection efficiency of a systemically delivered tumor-targeting immunolipoplex by inclusion of a pH-sensitive histidylated oligolysine peptide. *Nucleic Acids Res.* **32**:e48.
13. Hoekstra, D., Rejman, J., Wasungu, L., Shi, W.F., and Zuhorn, I. (2007) Gene delivery by cationic lipids: in and out of an endosome. *Biochemical Society Trans.* **35**:68–71.
14. Kukowska-Latallo, J.F., Bielinska, A.U., Johnson, J., Spindler, R., Tomalia, D.A., and Baker, Jr., J.R. (1996) Efficient transfer of genetic material into mammalian cells using Starburst polyamidoamine dendrimers *PNAS.* **93**:4897–4902.
15. Elbashir, S.M., Harborth, J., Lendeckel, W., Yalcin, A., Weber, K., and Tuschl, T. (2001) Duplexes of 21-nucleotide RNAs mediate RNA interference in cultured mammalian cells. *Nature.* **411**:494–4565.
16. Gregory, R.I., and Shiekhattar, R. (2005) MicroRNA biogenesis and cancer. *Cancer Res.* **65**:3509–3512.

17. Vermeulen, A., Behlen, L., Reynolds, A., Wolfson, A., Marshall, W.S., Karpilow, J., and Khvorova, A. (2005) The contributions of dsRNA structure to Dicer specificity and efficiency. *RNA*. **11**:674–682.
18. Reynolds, A., Leake, D., Boese, Q., Scaringe, S., Marshall, W.S., and Khvorova, A. (2004) Rational siRNA design for RNA interference. *Nat Biotechnol.* **22**, 326–330.
19. Tuzmen, S., Kiefer, J., and Mousses, S. (2007) Validation of short interfering RNA knockdowns by quantitative real-time PCR. *Methods Mol Biol.* **353**:177–203.
20. Brazas, R.M., and Hagstrom, J.E. (2005) Delivery of small interfering RNA to mammalian cells in culture by using cationic lipid/polymer-based transfection reagents. *Methods Enzymol.* **392**:112–124.
21. Persengiev, S.P., Zhu, X., and Green, M.R. (2004) Nonspecific, concentration-dependent stimulation and repression of mammalian gene expression by small interfering RNAs. *RNA*. **10**:12–18.

4

Poly(β-amino esters): Procedures for Synthesis and Gene Delivery

Jordan J. Green, Gregory T. Zugates, Robert Langer, and Daniel G. Anderson

Summary

Non-viral gene delivery systems are promising as they avoid many problems of viral gene therapy by having increased design flexibility, high safety, large DNA cargo capacity, and ease of manufacture. Here, we describe the use of polymeric vectors, in particular biodegradable poly(β-amino esters) (PBAEs), for non-viral gene delivery. These polymers are able to self-assemble with DNA and form positively charged gene delivery nanoparticles. Methods for synthesis of these polymers, particle self-assembly, and transfection using these particles are delineated. A standard protocol is presented as well as a high-throughput screening technique that can be used to more quickly optimize transfection parameters for efficient delivery.

Key Words: Gene therapy; Drug delivery; Polymers; Transfection; High-throughput screening.

1. Introduction

Gene delivery is a valuable research tool and also has strong therapeutic potential to treat many human diseases, from monogenic diseases to cancer. However, this potential has not yet been fully realized because a safe and efficient method for gene delivery has not yet been developed. Viral vectors have several safety concerns (e.g., immunogenicity, reversion to the wild type), limited DNA carrying capacity, and production/quality control challenges *(1,2)*. Engineered gene delivery biomaterials have increasingly been shown to address these concerns, but have lower efficiency than viruses *(3)*. Many biomaterials

From: *Methods in Molecular Biology, vol. 480: Macromolecular Drug Delivery*, Edited by: M. Belting
DOI 10.1007/978-1-59745-429-2_4, © Humana Press, a part of Springer Science+Business Media, LLC 2009

have been used for gene delivery including cationic polymers, liposomes, dendrimers, chitosans, and inorganic nanoparticles *(1,4)*. Cationic polymers, in particular, have been shown to be a flexible system to condense DNA into nanoparticles that are effective for gene delivery *(5)*.

A specific class of cationic polymers, poly(β-amino esters) (PBAEs), are promising as they condense DNA into nanoparticles, are biodegradable via hydrolytic cleavage of their ester groups, and have lower cytotoxicity compared to other cationic polymers such as polyethylenimine (PEI) *(6–8)*. These polymers may act through the "proton sponge mechanism" to enable escape from the endosomal compartment to the cytoplasm *(9–11)*. PBAEs have also been developed as gene delivery systems for the in vivo treatment of prostate cancer *(12)*, as targeted delivery systems using electrostatic coatings of peptide ligands *(13)*, and as transfection agents that rival the efficacy of viral gene delivery systems *(14)*.

Methods to transfect new cell types or quantify the transfection efficiency of novel biomaterials are fundamental to the development of polymeric gene delivery vectors. This chapter discusses high-throughput methods for screening new gene delivery biomaterials while varying important design parameters such as polymer structure, DNA loading basis, and polymer to DNA weight ratio in parallel. These approaches can be used to quickly characterize and optimize new gene delivery formulations.

2. Materials

2.1. Polymer Synthesis

1. For synthesis of one lead polymer, poly(5-amino-1-pentanol-*co*-1,4-butanediol diacrylate) (referred to as C32), 5-amino-1-pentanol (Aldrich) and 1,4-butanediol diacrylate (Scientific Polymer Products, Inc.) are required. Additional amine and diacrylate monomers can be purchased and used following the same protocol to create structurally diverse polymers that are useful for gene delivery. These monomers can be purchased from Aldrich (Milwaukee, WI, USA), Scientific Polymer Products, Inc. (Ontario, NY, USA), TCI (Portland, OR, USA), Pfaltz & Bauer (Waterbury, CT, USA), Matrix Scientific (Columbia, SC, USA), and Dajac Monomer–Polymer (Feasterville, PA, USA).
2. Teflon-lined screw cap glass vials, 8 mL; VWR
3. Teflon-coated magnetic micro stir bars that fit in vials; VWR
4. Magnetic stir plate
5. Incubator/oven (95°C)

2.2. Cell Culture

2.2.1. COS-7 Cells

1. COS-7 cells, African green monkey kidney fibroblast-like cell line; ATCC #CRL-1651

2. Medium: 500 mL Dulbecco's Modified Eagle's Medium (DMEM) containing 10% fetal bovine serum (50 mL FBS) and 1% Penicillin–Streptomycin (5 mL); Invitrogen
3. Trypsin; Invitrogen
4. Phosphate Buffered Saline (PBS); Invitrogen

2.2.2. Human Umbilical Vein Endothelial Cells (HUVECs)

1. HUVECs, primary human endothelial cells; Lonza #C2519A
2. Medium: 500 mL Endothelial Cell Growth Medium-2 (EGM-2) medium supplemented with SingleQuot kit; Lonza
3. ReagentPack kit (Trypsin/EDTA, Trypsin Neutralizing Solution, HEPES Buffered Saline); Lonza

2.2.3. Additional Materials

1. Hemocytometer; VWR
2. Incubator (37°C, 5% CO_2); Forma Scientific

2.3. Standard Gene Delivery Transfection

1. Gene delivery polymer(s). From synthesis in **Subheading 2.1.** and/or bought commercially such as PEI, $M_w \sim 25$ kDa branched from Sigma or PBAEs can be bought directly from Open Biosystems as Leopard™ Transfection Array polymers.
2. Dimethyl sulfoxide (DMSO), >99.7% and sterile; Sigma
3. Tissue culture filter unit, 500 mL, 0.2-μ cellulose acetate, sterile; Nalgene
4. Sodium acetate solution (3 M), pH 5.2, 0.2 μm filtered; Sigma
5. pEGFP DNA in water (1 mg/mL), stored at −20°C; Elim Biopharmaceuticals
6. Single channel pipettes (Ex: 1–10 μL, 10–100 μL, and 100–1000 μL)
7. Eppendorf tubes, 1.5 mL, sterile; VWR
8. Pipette tips (sterilized by autoclaving for 30 min at 18 psi and at 120°C)
9. Six-channel aspirator wand (autoclaved); V&P Scientific
10. Clear, sterile, tissue culture treated multi-well plates (6-well, 12-well, or 24-well, etc.)
11. Vortex-Genie; VWR
12. Fluorescent microscope and/or flow cytometer to measure green fluorescent protein (GFP) gene expression.

2.4. High-Throughput Gene Delivery Transfection

1. Gene delivery polymer(s). These could include materials derived from the synthesis as described in **Subheading 2.1.** and/or bought commercially, such as PEI, $M_w \sim 25$ kDa branched from Sigma, or PBAEs bought directly from Open Biosystems as Leopard™ Transfection Array polymers.
2. DMSO, >99.7% and sterile; Sigma
3. Tissue culture filter unit, 500 mL, 0.2 μ cellulose acetate, sterile; Nalgene

4. Sodium acetate solution (3 M), pH 5.2, 0.2 μm filtered; Sigma
5. pCMV-Luc DNA in water (1 mg/mL), stored at −20°C; Elim Biopharmaceuticals
6. Single channel pipettes (Ex: 1–10 μL, 10–100 μL, and 100–1000 μL)
7. Eppendorf tubes, 1.5 mL, sterile; VWR
8. Twelve-channel pipettes (5–50 μL and 50–300 μL); Finnpipette
9. Pipette tips (sterilized by autoclaving for 30 min at 18 psi and at 120°C)
10. Pipetting reservoirs, sterile; VWR
11. Twelve-channel aspirator wand (autoclaved); V&P Scientific
12. Clear 96-well half-area plate (sterilized by UV treatment in cell culture hood for at least 1 h); Corning #3695
13. White, opaque, sterile, tissue culture treated 96-well plate; Costar #3917
14. Clear 96-well flat-bottom plates with lids, sterile; BD Falcon #353072
15. Polypropylene 96-well plates (2.4 mL, V-bottom); Sigma #M1561
16. Bright-Glo™ kits; Promega
17. Firefly luciferase protein; Promega
18. Ninety-six-well plate luminometer to measure luciferase gene expression

3. Methods

3.1. Polymer Synthesis

1. Weigh 400 mg of 5-amino-1-pentanol (or other amine monomer) into a 5 mL sample vial with a teflon-lined screw cap.
2. Add acrylate monomer to amine monomer at an amine/diacrylate stoichiometric ratio of 1.2:1. For polymer C32, add 640 mg of 1,4-butanediol diacrylate to the 400 mg of 5-amino-1-pentanol.
3. Add a small teflon-coated magnetic stir bar to the vial.
4. To polymerize, stir the monomers on a magnetic stir plate in an oven at 95°C for 12 h (**Fig. 1**).
5. Remove polymer vial and store in the dark at 4°C until ready to use.

3.2. Standard Transfection

Researchers should be familiar with basic sterile cell culture techniques, the ability to grow, passage, and plate cells. All work is conducted in a laminar flow biosafety cabinet using sterilized reagents and equipment.

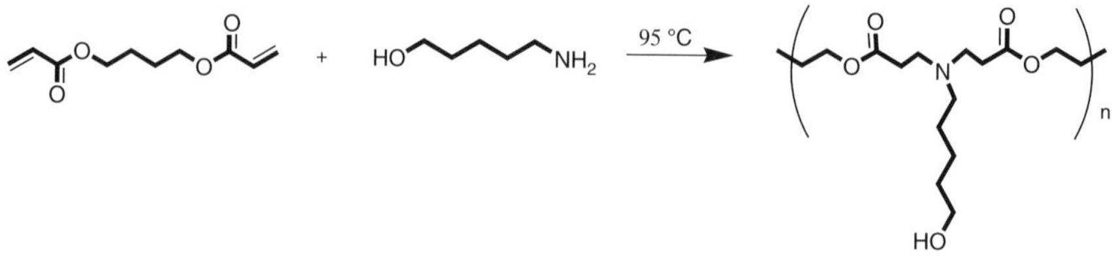

Fig. 1. Polymerization of 1,4-butanediol diacrylate and 5-amino-1-pentanol to form gene delivery polymer C32.

Table 1
Cell Plating

Wells/plate	Volume/well	Cells/well
6	2 mL	300,000
12	1 mL	150,000
24	500 µL	75,000
96	100 µL	15,000

1. **Cell plating.** Twenty-four hours prior to transfection, seed cells into a clear tissue culture multi-well plate at 150 cells/µL according to **Table 1**.
2. **Preparation of polymer stock solutions.** Dissolve 50 mg of each synthesized polymer in 500 µL of DMSO in a sterilized eppendorf tube. Vortex to mix to prepare 100 mg/mL polymer stock solutions. For PEI, dissolve 1 mg in 1 mL of water to prepare a 1 mg/mL PEI stock solution.
3. **Preparation of sodium acetate buffer.** To prepare sodium acetate (NaAc) buffer (pH 5.2), dilute 4.2 mL of 3 M sodium acetate into 495.8 mL of deionized water. Sterilize by vacuum filtration through a 0.2 µm filter.
4. **Change cell media.** On the day of transfection, warm fresh media in a 37°C water bath. Aspirate out the old media from each cell well and replace with equivalent volume of fresh media.

    Note: The cells should look healthy as determined by light microscopy. Transfections are typically performed at 70–100% confluence.
5. **DNA preparation.** Thaw the DNA stock solution at room temperature. Dilute an aliquot of the 1 mg/mL DNA stock solution with 25 mM sodium acetate buffer to a final concentration of 0.060 mg/mL. For a 24-well plate, this requires 90 µg of DNA in a final volume of 1.5 mL of 25 mM sodium acetate. Aliquot out 60 µL of diluted DNA into a small volume sterile eppendorf tube for each sample to be made (24 typically).
6. **Aqueous polymer preparation.** Prior to complexation with DNA, each 100 mg/mL polymer/DMSO solution must be diluted in sodium acetate. The dilution is dependent on the final polymer:DNA weight ratios desired. For polymer C32, a weight ratio (w/w) of 30:1 polymer:DNA is typically optimal. However, this may vary with different polymers and different cell systems. Generally, it is advisable to initially try a range of weight ratios and then select the best-performing w/w for future experiments. The aqueous polymer solution should be vigorously mixed prior to use to ensure homogeneity. **Table 2** shows a protocol for typical formulations in a 24-well plate format assuming polymer preparation for triplicate samples (200 µL aqueous polymer solution; for different plate formats, all volumes are scaled in proportion to the cell seeding density used in **step 1**). PEI is prepared analogously except that a low 1:1 polymer to DNA weight ratio is generally optimal and the PEI (and the DNA used to complex PEI) is diluted in 150 mM sodium chloride rather than 25 mM sodium acetate.

Table 2
Polymer Preparation

Polymer: DNA (w/w)	Polymer (μg)	Polymer/ DMSO (μL)	NaAc (μL)	Polymer concentration (μg/μL)
1	12	1.2	199	0.06
5	60	6.0	194	0.30
10	120	12	188	0.60
20	240	24	176	1.20
30	360	36	164	1.80
50	600	60	140	3.00
100	1200	120	80	6.00

7. **Polymer–DNA complexation/nanoparticle formation.** For each polymer replicate sample, add 60 μL of polymer to an eppendorf tube containing 60 μL DNA. Mix by vortexing on a medium setting for 10 s. Time 10 min on a timer to allow for polymer/DNA self-assembly prior to use.

8. **Add polymer/DNA nanoparticles to the cells.** Prior to addition of the polymer/DNA particles, the media over the seeded cells may be optionally removed and replaced with new media with altered composition for the transfection (serum-free or high-serum media for example), although this is not required. Polymeric gene delivery particles are then added to each well according to **Table 3**. When finished, swirl the plate, return the cells to the incubator, and start a timer.

9. **Remove polymer/DNA particles from the cells.** Typically, particles are incubated on the cells for 1–4 h. Depending on the concentration and the toxicity of materials, in some case overnight incubations are helpful. After incubation, aspirate the polymer/DNA particles from each well using either a 6-channel aspirating wand or a Pasteur pipette. Add a volume of fresh, warm media equal to the initial cell seeding volume used in each well (i.e., 500 μL for a 24-well plate). Return the cells to the incubator.

10. **GFP expression measurement.** Alternate strategies may be used to quantify GFP gene expression. Initial study can be performed using a fluorescent microscope to measure the percentage of green, expressing cells vs. total cells counted

Table 3
Gene Delivery Nanoparticle Dose

Wells/plate	Volume/well	Particle volume/well (μL)	DNA/well (μg)
6	2 mL	400	12
12	1 mL	200	6
24	500 μL	100	3
96	100 μL	20	0.6

in the bright field image. Typically, however, fluorescent activated cell sorting (FACS) is used to quickly count 10,000 or more live cells per sample and gate the GFP positive cells from the GFP negative cells. For analysis of GFP positive cells, it is recommended that two-dimensional gating rather than a one-dimensional histogram is used. For two-dimensional gating, the green channel is plotted on the *x*-axis and the yellow channel is plotted on the *y*-axis. In this manner, GFP positive cells can be better differentiated from increased background auto-fluorescence as shown in **Fig. 2**.

3.3. High-Throughput Screening (15)

The following protocol is for a 96-well plate testing 11 different polymers (columns 1–11 are for each polymer and column 12 is for control). Each polymer is tested in quadruplicate at two different polymer to DNA w/w ratios (20 w/w is tested in the top-half of the plate (rows A–D) and 100 w/w is tested in the bottom-half of the plate (rows E–H)).

1. **Cell plating.** Twenty-four hours prior to transfection, seed 15,000 cells/well into a white 96-well tissue culture plate.
2. **Preparation of polymer stock solutions.** Dissolve 50 mg of each synthesized polymer in 500 µL of DMSO in a sterilized eppendorf tube. Vortex to mix to prepare 100 mg/mL polymer stock solutions. For PEI, dissolve 1 mg in 1 mL of water to prepare a 1 mg/mL PEI stock solution.
3. **Preparation of sodium acetate buffer.** To prepare sodium acetate (NaAc) buffer (pH 5.2), dilute 4.2 mL of 3 *M* sodium acetate into 495.8 mL of deionized water. Sterilize by vacuum filtration through a 0.2 µm filter.

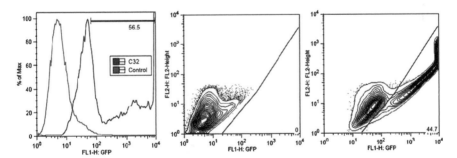

Fig. 2. Representative HUVEC transfection efficacy of C32 and method for FACS gating. The one-dimensional histogram of C32/DNA transfected cells (56.5% positive) (*left*) includes some falsely positive cells that are excluded during the two-dimensional analysis of the same data set as shown by the FACS density plot gating for the negative control (0% positive) (*middle*) and the C32/DNA transfection (44.7% positive) (*right*). Ratio of GFP fluorescence (*x*-axis) to background fluorescence (*y*-axis) is used to accurately gate positive cells.

4. **Plating sodium acetate buffer.** First, transfer 30 mL of sodium acetate buffer to a pipetting reservoir. Using a multi-channel pipette, add 900 μL/well NaAc to row A of a 2.4 mL deep 96-well plate (**Fig. 3**, plate #1). Next, in a clear 96-well plate, add 176 μL/well NaAc to row A (for 20 w/w polymer to DNA particles) and 80 μL/well NaAc to row B (for 100 w/w polymer to DNA particles) (**Fig. 3**, plate #2).

5. **DNA preparation and plating.** Thaw the pCMV-Luc DNA stock solution at room temperature. Dilute 600 μL of the 1 mg/mL DNA stock with 9.4 mL of sodium acetate buffer in a 15 mL sterile tube. Transfer the diluted DNA solution to a pipetting reservoir and add 25 μL/well to all wells of a 96-well half-area plate (**Fig. 3**, plate #3).

6. **Plating media.** Warm 25 mL of media in a 37°C water bath, transfer the media to a pipetting reservoir, and add 200 μL/well to each well of a new clear 96-well plate (**Fig. 3**, plate #4).

7. **Aqueous polymer preparation.** For each 100 mg/mL polymer/DMSO solution and the positive control, add 100 μL of concentrated polymer to a single well (#1–12) in row A of the 2.4 mL deep 96-well plate containing 900 μL of sodium

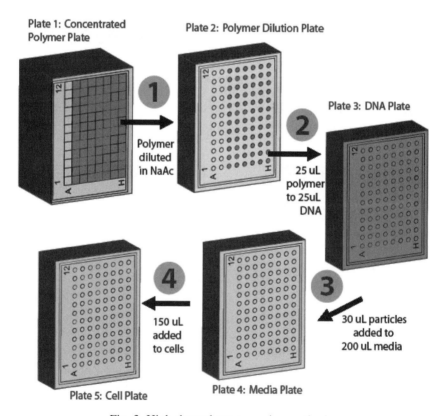

Plate 1: Concentrated Polymer Plate

Plate 2: Polymer Dilution Plate

Plate 3: DNA Plate

1 Polymer diluted in NaAc

2 25 uL polymer to 25uL DNA

3 30 uL particles added to 200 uL media

4 150 uL added to cells

Plate 5: Cell Plate

Plate 4: Media Plate

Fig. 3. High-throughput screening method.

acetate buffer (**Fig. 3**, plate #1). Vigorously pipette the solution several times to ensure that the polymers are dissolved fully in the buffer.

Note: For **steps #8–13**, all pipetting is done using 12-channel pipettes. Change pipette tips after each use.

8. **Polymer dilutions (Fig. 3, step 1).** Add 24 μL of polymer solution from row A of the 2.4 mL deep 96-well plate (plate #1) to row A of the polymer dilution plate (plate #2) containing 176 μL of sodium acetate buffer. Similarly, add 120 μL of polymer solution from row A of the 2.4 mL deep 96-well plate (plate #1) to row B of the polymer dilution plate (plate #2) containing 80 μL of sodium acetate buffer. Vigorously pipette the solutions multiple times to ensure that they are well mixed.

9. **Polymer–DNA complexation/nanoparticle formation(Fig. 3, step 2).** For quadruplicate samples, add 25 μL of polymer from row A of the polymer dilution plate (plate #2) to 25 μL of DNA in rows A–D of the DNA plate (plate #3). Add 25 μL of polymer from row B of the polymer dilution plate (plate #2) to 25 μL of DNA in rows E–H of the DNA plate (plate #3). For each addition, vigorously pipette the solution several times to promote polymer/DNA self-assembly. To generate reproducible particles, it is important to be consistent with the mixing technique. Start a 5 min timer once the plate is finished.

10. **Polymer/DNA particles dilution (Fig. 3, step 3).** At the end of 5 min, add 30 μL of polymer/DNA particles from row A of the DNA plate (plate #3) to row A of the media plate (plate #4) containing 200 μL of media. Repeat for rows B–H.

11. **Add polymer/DNA particles to the cells (Fig. 3, step 4).** Use a 12-channel aspiration wand to remove the media from the cells seeded into the white 96-well plate previously (**step #1**). Add 150 μL of polymer/DNA particles from row A of the media plate (plate #4) to row A of the cell plate. Repeat for rows B–H, being careful to pipette against the side of the wells as opposed to directly over top the cells which can dislodge the cells if not careful.

12. **Remove polymer/DNA particles from the cells.** After a 1–4 h incubation time, aspirate the polymer/DNA solution from the cells and add 100 μL of warm fresh media to each well. Return the cells to the incubator.

13. **Luciferase protein assay.** This assay is generally performed 2–3 days post-transfection. Thaw the Bright-Glo™ Luciferase Assay Kit by warming the buffer in a room temperature water bath. Remove the cells from the incubator and allow them to equilibrate to room temperature as well. Add the buffer to the substrate vial, re-cap, invert to mix, and then dispense into a reservoir. Using a multi-channel pipette, add 100 μL of the solution to each well of the cell plate. After the last addition, time 2 min on a timer. Place the plate on an orbital shaker or manually shake to promote mixing during this time. After 2 min, measure the luminescence in a plate reader using an integration time of 1 s/well. Compare the relative luminescence to a standard curve to obtain the mass of luciferase protein in each well.

4. Notes

1. In general, any adherent cell type may be used following these same protocols. COS-7 and HUVEC cell protocols are shown as examples of an easier-to-transfect mammalian cell line and more difficult-to-transfect human primary cells, respectively. Typical transfection of COS-7 cells using polymer C32 is ~90–100% whereas for HUVECs it is ~40–50%.
2. Different-sized plasmid DNA may be used to create these polymer/DNA particles. For easy visual inspection of gene expression and quantitative cell population information enhanced green fluorescent protein (EGFP) DNA is often used. For high-throughput screening applications, luciferase (Luc) DNA is often used instead.
3. It is important that the cationic polymers are added to the anionic DNA, rather than the reverse. It is also important that the two components are very well mixed. We suggest vortexing at medium speed rather than simply pipetting up and down. The optimal self-assembly incubation time is between 10 and 15 min. Leaving the assembled polymer/DNA particles for longer than 40 min before use may reduce activity. A slightly shorter self-assembly waiting time is typically used when performing high-throughput methods.
4. If for a given application transfection is lower than desired, increase the polymer to DNA weight ratio and/or the DNA loading basis per cell well. For example, in a 24-well plate use a 6 μg DNA/well basis instead of a 3 μg DNA/well basis. Additionally, increasing the time that the polymer/DNA particles incubate with the cells or reducing the serum concentration can increase efficacy. Lastly, reducing the cell confluence to 50–70% confluent at the time of transfection can increase efficacy.
5. If for a given application cytotoxicity occurs, decrease the polymer to DNA weight ratio and/or the DNA loading basis per cell well.
6. High-Throughput Screening Advantages/Limitations:

4.1. Advantages

- Allows for testing of hundreds of polymers, in quadruplicate, in a single day
- Minimizes amount of reagents used

4.2. Limitations

- Many pipette tip boxes and plate types must be stocked and sterilized
- Mixing, during parallel particle formation, by pipettes may not be as vigorous as done by vortexing. This can affect particle self-assembly in some instances.

References

1. Partridge, K. A., and Oreffo, R. O. C. (2004) Gene delivery in bone tissue engineering: Progress and prospects using viral and nonviral strategies. *Tissue Eng.* **10,** 295–307.

2. Thomas, C. E., Ehrhardt, A., and Kay, M. A. (2003) Progress and problems with the use of viral vectors for gene therapy. *Nat. Rev. Genet.* **4,** 346–358.
3. Pack, D. W., Hoffman, A. S., Pun, S., and Stayton, P. S. (2005) Design and development of polymers for gene delivery. *Nat. Rev. Drug Discov.* **4,** 581–593.
4. Merdan, T., Kopecek, J., and Kissel, T. (2002) Prospects for cationic polymers in gene and oligonucleotide therapy against cancer. *Adv. Drug Delivery Rev.* **54,** 715–758.
5. Putnam, D. (2006) Polymers for gene delivery across length scales. *Nat. Mater.* **5,** 439–451.
6. Lynn, D. M., and Langer, R. (2000) Degradable poly(beta-amino esters): Synthesis, characterization, and self-assembly with plasmid DNA. *J. Am. Chem. Soc.* **122,** 10761–10768.
7. Green, J. J., Shi, J., Chiu, E., Leshchiner, E. S., Langer, R., and Anderson, D. G. (2006) Biodegradable polymeric vectors for gene delivery to human endothelial cells. *Bioconjug. Chem.* **17,** 1162–1169.
8. Anderson, D. G., Lynn, D. M., and Langer, R. (2003) Semi-automated synthesis and screening of a large library of degradable cationic polymers for gene delivery. *Ang. Chem. Int. Edn* **42,** 3153–3158.
9. Akinc, A., and Langer, R. (2002) Measuring the pH environment of DNA delivered using nonviral vectors: Implications for lysosomal trafficking. *Biotechnol. Bioeng.* **78,** 503–508.
10. Sonawane, N. D., Szoka, F. C., and Verkman, A. S. (2003) Chloride accumulation and swelling in endosomes enhances DNA transfer by polyamine-DNA polyplexes. *J. Biol. Chem.* **278,** 44826–44831.
11. Akinc, A., Thomas, M., Klibanov, A. M., and Langer, R. (2005) Exploring polyethylenimine-mediated DNA transfection and the proton sponge hypothesis. *J. Gene Med.* **7,** 657–663.
12. Anderson, D. G., Peng, W. D., Akinc, A., Hossain, N., Kohn, A., Padera, R., Langer, R., and Sawicki, J. A. (2004) A polymer library approach to suicide gene therapy for cancer. *Proc. Natl. Acad. Sci. U.S.A.* **101,** 16028–16033.
13. Green, J. J., Chiu, E., Leshchiner, E. S., Shi, J., Langer, R., and Anderson, D. G. (2007) Electrostatic ligand coatings of nanoparticles enable ligand-specific gene delivery to human primary cells. *Nano Lett.* **7,** 874–879.
14. Green, J. J., Zugates, G. T., Tedford, N. C., Huang, Y., Griffith, L. G., Lauffenburger, D. A., Sawicki, J. A., Langer, R., and Anderson, D. G. (2007) Combinatorial modification of degradable polymers enables transfection of human cells comparable to adenovirus. *Adv. Mater.* **19**(19), 2836–2842.
15. Zugates, G. T., Anderson, D. G., and Langer, R. (2007) High-throughput methods for screening polymeric transfection reagents, in *Gene Transfer: Delivery and Expression of DNA and RNA* (Friedmann, T., and Rossi, J., eds), Cold Spring Harbor Laboratory Press, New York, NY, pp. 547–54.

5

Systemic Delivery and Pre-clinical Evaluation of Nanoparticles Containing Antisense Oligonucleotides and siRNAs

Chuanbo Zhang, Joseph T. Newsome, Rajshree Mewani, Jin Pei, Prafulla C. Gokhale, and Usha N. Kasid

Summary

By virtue of their potential to selectively silence oncogenic molecules in cancer cells, antisense oligonucleotides (ASO) and small interfering RNAs (siRNAs) are powerful tools for development of tailored anti-cancer drugs. The clinical benefit of ASO/siRNA therapeutic is, however, hampered due to poor pharmacokinetics and biodistribution, and suboptimal suppression of the target in tumor tissues. Raf-1 protein serine/threonine kinase is a druggable signaling molecule in cancer therapy. Our laboratory has developed cationic liposomes for systemic delivery of raf ASO (LErafAON) and raf siRNA (LErafsiRNA) to human tumor xenografts grown in athymic mice. LErafAON is also the first ASO containing liposomal drug tested in humans. In this article, we primarily focus on a modified formulation of systemically delivered cationic liposomes containing *raf* antisense oligonucleotide (md-LErafAON). The cationic liposomes were prepared using dimyristoyl 1,2-diacyl-3-trimethylammonium-propane (DMTAP), phosphatidylcholine (PC), and cholesterol (CHOL). The toxicology, pharmacokinetics, biodistribution, target selectivity, and anti-tumor efficacy studies of md-LErafAON were conducted in mice. We demonstrate that md-LErafAON is the next generation of systemically delivered and well-tolerated antisense therapeutic suitable for clinical evaluation.

Key Words: Raf-1; Antisense oligonucleotide; siRNA; Cationic liposomes; Dimyristoyl 1,2-diacyl-3-trimethylammonium-propane (DMTAP); Systemic delivery; Toxicology; Pharmacokinetics and biodistribution; Ionizing radiation; Prostate cancer.

From: *Methods in Molecular Biology, vol. 480: Macromolecular Drug Delivery,* Edited by: M. Belting
DOI 10.1007/978-1-59745-429-2_5, © Humana Press, a part of Springer Science+Business Media, LLC 2009

1. Introduction

It has been well established that oncogene-specific antisense oligonucleotides (ASO) and small interfering RNAs (siRNAs) are able to knock down the target protein in cancer cells and these reagents hold tremendous promise for the design and development of tailored cancer therapies *(1–3)*. The application of ASO/siRNA therapeutic is, however, limited due to serum nuclease degradation and poor cellular uptake. Numerous reports have demonstrated that the modified versions of certain ASO/siRNAs are viable in vivo *(4–10)*. In addition, systemically administered nanoparticles carrying ASO/siRNA have been shown to effectively inhibit the target molecules in tumor tissues in mouse models *(11–21)*.

We and others have demonstrated that Raf-1 protein serine/threonine kinase is a druggable signaling molecule in cancer therapy *(1,13,17,21–25)*. Our laboratory has developed a novel cationic liposomal formulation for systemic delivery of intact raf ASO (LErafAON) to normal and tumor tissues in mice *(13,17)*. The liposome-entrapped raf antisense oligonucleotide (LErafAON) is also the first liposomal ASO drug tested in humans *(26,27)*. Systemically delivered cationic liposomal nanoparticles containing rafsiRNA (LErafsiRNA) also inhibit Raf-1 protein expression in tumor and most normal tissues in human prostate tumor (PC-3)-bearing athymic mice (**Fig. 1** and Color Plate 1, *see* Color Plate Section).

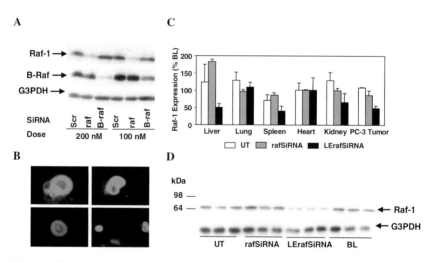

Fig. 1. Effects of systemically delivered LErafsiRNA on Raf-1 expression in normal and tumor tissues. (**a**) Validation of target specificity of rafsiRNA in vitro. Human prostate carcinoma cells (PC-3) were treated with indicated concentration of the siRNA [Raf-1, B-Raf, or Scrambled (Scr)] and Oligofectamine according to supplier's instructions (Invitrogen Corp., Carlsbad, CA, USA) as described earlier *(28)*. After 72 h,

Fig. 1. (Continued) the medium was discarded and the plate was rinsed twice with 1 × PBS (Invitrogen Corp.), followed by the addition of 200μL of lysis buffer (1% Triton X-100, 0.1% SDS, 0.5% sodium deoxycholate, 10mM NaCl containing protease inhibitor cocktail). The whole cell lysates were analyzed by 12% SDS–Tris–glycine PAGE and by Western blotting using polyclonal anti-Raf-1 antibody (C-12) (Santa Cruz Biotechnology, GA, USA). The same blot was sequentially reprobed with polyclonal B-Raf antibody (Santa Cruz Biotechnology, GA, USA), and polyclonal anti-G3PDH antibody (Trevigen, Inc., Gaithersburg, MD). Signals were quantified using the NIH Image J 1.31 software program (rsb.info.nih.gov/ij/). (**b**) Entrapment of rafsiRNA in the lipid bilayer of LErafsiRNA. 5′-Fluorescein-labeled Raf-1 siRNA (FITCrafsiRNA) was custom ordered (fluorescein-5′-UGU GCG AAA UGG AAU GAG C [dT] [dT]-3′ (Invitrogen Corp.)). The GMPs grade formulation of lyophilized lipids containing a mixture of dimethyldioctadecylammonium bromide (DDAB), PC, and CHOL in a molar ratio 1:3.2:1.6 was prepared, as we have described earlier *(13)*. FITCrafsiRNA was reconstituted in normal saline at a concentration of 276.7 μg/mL. Liposome-entrapped FITCrafsiRNA (LEFITCrafsiRNA) was prepared by hydrating 375 μg of lyophilized lipids, containing 64.2 μg DDAB, 247.8 μg PC, and 63 μg CHOL, with the FITCraf-siRNA solution (25 μg of rafsiRNA) for 2 h at room temperature. The FITCrafsiRNA to lipid ratio was 1:15 (w/w). The mixture was vortexed gently, followed by sonication for 10 min at room temperature in a bath type sonicator (Model XL 2020, Misonix, Inc., Farmingdale, NY, USA). The fluorescent images were taken using Olympus Fluoview FV300 Laser Scanning Confocal microscope. (**c**) Quantification data showing the effects of LErafsiRNA on Raf-1 expression in normal and tumor tissues of athymic mice systemically treated with LErafsiRNA nanoparticle. Logarithmically growing PC-3 cells (5 × 10^6 cells) were injected s.c. into the left flank region of male athymic NCR nu/nu mice (National Cancer Institute). Tumors were allowed to grow to a mean tumor volume of ~100 mm^3. Tumor-bearing mice were randomly divided into four treatment groups ($n = 3$). LErafsiRNA was formulated as described in panel B. LErafsiRNA or rafsiRNA was administered i.v. via tail vein (70 μg/dose, × 4, over 48 h), and mice were sacrificed 24 h after the last injection. Control mice received BL at the same dosing and schedule as LErafsiRNA or were left untreated (UT). Tumor, kidneys, spleen, liver, lungs, and heart were excised and stored at −80°C. Various tissues were homogenized in cell lysis buffer, prepared as in panel A, using Omni tissue homogenizer (Omni International, GA, USA). The whole cell extracts (50 μg protein) were analyzed by Western blotting using anti-Raf-1 antibody. The blots were probed with anti-GAPDH antibody, and the GAPDH normalized data were used to quantify the relative expression of Raf-1. (**d**) Representative Western blot showing expression of Raf-1 (~75 kDa) and G3PDH (loading control) in spleen of mice in various treatment groups. All siRNA duplex sequences were custom ordered as desalted grade from Invitrogen Corp. Raf-1 sense; 5′-UGU GCG AAA UGG AAU GAG C [dT] [dT]-3′; B-Raf sense; 5′-AAG UGG CAU GGU GAU GUG GCA-3′, and Scrambled sense; 5′-AAG UCC AUG GUG ACA GGA GAC-3′. The particle size of LErafsiRNA was 235 ± 7.6 nm (mean ± SD, $n = 3$) (*See* Color Plate 1, *see* Color Plate Section).

Further modifications of the liposomal composition have led to an improved formulation of the cationic liposomes containing raf ASO (md-LErafAON). The md-LErafAON formulation exhibits superior safety and pharmacokinetic profiles and is an effective anti-tumor agent in a human prostate tumor xenograft model.

In this chapter, we describe four major protocols: formulation of md-LErafAON, toxicology of systemically delivered md-LErafAON, pharmacokinetics and biodistribution of md-LErafAON, and anti-tumor efficacy of a combination of md-LErafAON and radiation. Representative examples of experimental data with detailed legends are provided to further clarify the methods described. Additional information, variations, and alternatives to the methods are described under **Heading 4** under each protocol. For further information, the reader is referred to the earlier reports *(13–15,17,26,27)*.

2. Materials

2.1. Formulation of md-LErafAON

1. Antisense oligodeoxyribonucleotide sequence (5′-GTG CTC CAT TGA TGC-3′) directed toward the translation initiation site of human c-*raf*-1 mRNA (rafAON), complementary sense strand sequence (5′-GCA TCA ATG GAG CAC-3′; raf-SON), and mismatch sequence (5′-GTG TTC GAC CTA TGC-3′; MM) is custom synthesized at Hybridon Specialty Products (Milford, MA, USA) (*see* **Notes 1–3**).
2. Dimyristoyl 1,2-diacyl-3-trimethylammonium-propane (DMTAP), eggphosphatidylcholine (PC), and cholesterol (CHOL) are purchased from Avanti Polar Lipids (Alabaster, AL, USA).

Additional materials are described as needed throughout the text. Important equipment information is provided in **Subheading 3.1**.

2.2. Toxicology of Systemically Delivered md-LErafAON

1. Materials needed are described throughout the text, or in references cited. Equipment information is provided in **Heading 3**. All chemicals are of reagent grade.
2. Animals: Male CD2F1 mice, 5 weeks old, 20–22 g can be purchased from National Cancer Institute, NIH (Frederick, MD, USA). Mice were maintained in the AAALAC accredited Research Resources Facility of the Division of Comparative Medicine, Georgetown University Medical Center, and fed Purina chow and water ad libitum.

2.3. Plasma Pharmacokinetics and Biodistribution of md-LErafAON

1. Human prostate carcinoma cells (PC-3).
2. Improved Minimum Essential Medium (IMEM; Biofluids, Inc., Rockville, MD, USA) supplemented with 10% heat-inactivated fetal bovine serum, 100 µg/mL streptomycin, 100U/mL penicillin, and 2m*M* glutamine.

3. rafAON size markers are custom made by Gibco BRL (Rockville, MD, USA).

4. T4 polynucleotide kinase (New England Biolabs, Beverly, MA, USA); phenol–chloroform (Gibco BRL, Rockville, MD, USA); γ^{32}P-ATP (NEN, Boston, MA, USA); Chroma Spin-10 columns (Clontech, Palo Alto, CA, USA); ExpressHyb^{TM} Hybridization Solution (Clontech, Palo Alto, CA, USA); 20 × saline–sodium citrate (SSC); and 10% sodium dodecyl sulfate (SDS; Biofluids, Rockville, MD).

5. Reagents: 40% acrylamide/bis (19:1) solution and nylon membrane (Bio-Rad, Hercules, CA, USA); urea (Sigma, St. Louis, MO, USA); tetramethylethylenediamine (TEMED), ammonium persulfate, and 10 × Tris/Borate/EDTA (TBE) buffer (Gibco BRL, Rockville, MD, USA); SDS-polyacrylamide gel (PAGE) system (Gibco BRL, Rockville, MD, USA); 10 × Tris/Glycine/SDS buffer solution, 10 × Tris/Glycine buffer solution, Laemmli sample buffer, and 2-mercaptoethanol (Bio-Rad, Hercules, CA, USA).

6. Anti-G3PDH rabbit polyclonal antibody (Trevigen, Gaithersburg, MD, USA); Horseradish peroxidase linked anti-mouse or anti-rabbit immunoglobin (Amersham Pharmacia Biotech, Piscataway, NJ, USA); Triton X-100 (Bio-Rad, Hercules, CA, USA); aprotinin and leupeptin (Roche, Mannheim, Germany).

7. Vertical electrophoresis systems (Amersham Pharmacia Biotech, Piscataway, NJ, USA); *Trans*-Blot semi-dry transfer cell (Bio-Rad, Hercules, CA, USA); hybridization incubator (Fisher, Hanover Park, IL, USA); personal densitometer SI (Molecular Dynamics, Sunnyvale, CA, USA). Additional materials and equipment needed are described throughout the text. All chemicals are of reagent grade.

8. Animals: Male Balb/c nu/nu mice, 6–8 weeks old, can be purchased from National Cancer Institute, NIH (Frederick, MD, USA). Mice were maintained in the AAALAC accredited Research Resources Facility of the Division of Comparative Medicine, Georgetown University Medical Center, and fed Purina chow and water ad libitum.

2.4. Therapeutic Efficacy of a Combination of Systemically Delivered md-LErafAON and Ionizing Radiation (IR)

1. Materials needed are described throughout the text. Equipment information is provided in **Heading 3**. All chemicals are of reagent grade.

2. Animals: Male Balb/c nu/nu mice, 6–8 weeks old, can be purchased from National Cancer Institute, NIH (Frederick, MD, USA). Mice were maintained in the AAALAC accredited Research Resources Facility of the Division of Comparative Medicine, Georgetown University Medical Center, and fed Purina chow and water ad libitum.

3. Methods
3.1. Formulation of md-LErafAON

1. We describe here, the lyophilization method for the preparation of md-LErafAON (*see* **Note 4**). Lipids (5 mg DMTAP, 20 mg PC, and 5 mg CHOL) are dissolved in 4 mL

t-butanol and lyophilized using LyoStar Freeze Dryer (FTSSystems, Inc., Stone Ridge, NY, USA). The lyophilized lipids (30 mg) are reconstituted at room temperature in a vial containing 1mL of 2 mg/mL rafAON in normal saline (rafAON to lipid ratio, 1:15 w/w). The mixture is vortexed vigorously for 2 min and hydrated at room temperature for 2 h (*see* **Note 5**). At the end of hydration, mixture is sonicated for 10 min in a bath type sonicator (Model XL 2020, Misonix, Inc., Farmingdale, NY, USA) to achieve a homogenous solution, thus the md-LErafAON. Liposome-entrapped mismatch oligo (md-LEMM) is prepared as above. Blank liposomes (BL) are prepared exactly as described above in the absence of rafAON (*see* **Note 6**).

2. The liposome-entrapment efficiency of rafAON in md-LErafAON is determined by the denaturing gel electrophoresis method. First an aliquot of the md-LErafAON is designated as the pre-wash sample. The unentrapped rafAON is removed from the remainder of the preparation by first washing the liposomes twice in normal saline, followed by ultracentrifugation at 100,000 g for 20 min (post-wash sample). RafAON is extracted from pre- and post-wash samples with phenol–chloroform and the extracts are electrophoresed on a 20% polyacrylamide/8*M* urea gel, followed by electroblotting onto a nylon membrane in 0.5 × TBE buffer at 20 V for 1 h. The blot is probed with ^{32}P-end-labeled complementary rafSON probe in ExpressHyb Hybridization Solution (Clontech, Palo Alto, CA, USA). The radiolabeled-rafSON is generated by the 5′-end labeling with γ^{32}P-ATP using T4 polynucleotide kinase and purification over Chroma Spin-10 columns (Clontech, Palo Alto, CA, USA). The autoradiographs are scanned and the rafAON band (15-mer) quantified using ImageQuant software (Personal Densitometer, Molecular Dynamics, Sunnyville, CA, USA). The percent-entrapment efficiency of rafAON in liposomes is calculated by the formula: (post-wash signal/pre-wash signal) × 100.

As an example, the liposome-entrapment efficiency of md-LErafAON formulation was found to be greater than 80% in two independent experiments (*see* **Note 7**).

3.2. Toxicology of Systemically Delivered md-LErafAON

3.2.1. Experimental Design and Treatment Conditions

Male CD2F1 mice receive a total of 12 intravenous (i.v.) injections of md-LErafAON at a dose of 5, 15, 25, 35, 50, or 70 mg/kg of md-LErafAON or BL (*n* = 5, animals in each cohort are identified by the letters A–E). The md-LErafAON preparation is administered via tail vein on days 0, 1, 3, 4, 6, 7, 9, 10, 12, 13, 15, and 16. Control groups receive BL or are left untreated. All animals are observed at least once daily for any clinical signs of illness and weighed as a group on days 0, 3, 6, 9, 12, 15, 18, 21, 25, and 31. Individual animal weights are measured prior to necropsy. Two mice (identified by the letters A and B) from each treatment group are scheduled-sacrificed on day 18, representing short-term toxicity response. The remaining three mice (identified by the letters C–E) from each treatment group are scheduled-sacrificed on day 31, representing long-term toxicity response.

3.2.2. Blood Collection

Individual blood samples are collected from all animals in various groups at the time of sacrifice. Blood is obtained via intracardiac stick under methoxyflurane anesthesia and collected into EDTA microtainers and Serum Separator microtainers for complete blood count (CBC) and serum chemistry evaluation, respectively. Clinical chemistry is performed by a commercial resource (e.g., ANTECH Diagnostics) (*see* **Note 8**).

3.2.3. Gross Necropsy and Histopathology

Prosection is done and gross findings are recorded on necropsy sheets. Organs (liver, spleen, kidneys, heart, and lungs) from euthanatized animals are collected in 10% neutral buffered formalin (NBF) and submitted for hematoxylin and eosin (H&E) staining and histopathological evaluation to a commercial resource (e.g., Pathology Associates, Inc.). Gross and histopathological findings are graded by a board-certified veterinary pathologist as minimal, slight, moderate, or severe based on the degree of involvement relative to the untreated control.

3.2.4. Statistical Analysis

Student's *t*-test is performed to determine the significance of changes in clinical pathology values obtained in CD2F1 mice treated with md-LErafAON and control groups. Analysis includes both short-term and long-term treatment groups of these mice.

As an example, systemically administered md-LErafAON was found to be very well-tolerated in CD2F1 mice (**Fig. 2**).

3.3. Plasma Pharmacokinetics and Biodistribution of md-LErafAON

3.3.1. Cell Culture

Human prostate carcinoma cells (PC-3) are grown as a monolayer in IMEM supplemented with 10% heat-inactivated fetal bovine serum, 100 µg/mL streptomycin, 100U/mL penicillin, and 2mM glutamine.

3.3.2. Tumor Xenografts in Mice, Treatments, and Sample Collection

Logarithmically growing PC-3 cells are injected subcutaneously (s.c.) (5×10^6 cells) on the back of male Balb/c nu/nu mice. When the tumors grow to a mean

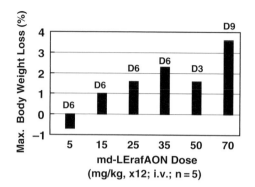

Fig. 2. md-LErafAON is non-toxic in CD2F1 mice. Mice received a total of 12 i.v. injections of md-LErafAON at a dose of 5, 15, 25, 35, 50, or 70 mg/kg administered over 17days (day 0, 1, 3, 4, 6, 7, 9, 10, 12, 13, 15, 16; $n = 5$). The cumulative dose of md-LErafAON administered in each of these groups was 60, 180, 300, 420, 600, or 840 mg/kg, respectively ($n = 5$). Control groups received BL at md-LErafAON equivalent lipid doses or were left untreated ($n = 5$). Group body weights were monitored at least twice a week. Two mice in each group were sacrificed on day 18 (short-term toxicity response), and the remaining three mice in each group were sacrificed on day 31 post-treatment initiation (long-term response). No clinical toxicity or death was noticed in any of the groups tested. Mice in untreated control group gained weight throughout the duration of study with a maximum weight gain of 18.5% at the time of sacrifice (day 31, post-treatment initiation). A transient, clinically insignificant maximum group body weight loss of less than 4% (day 9) was observed in the highest dose group (70 mg/kg). All experimental groups showed steady weight gain, during the course of treatment, after day 6 (5–35 mg/kg groups), day 3 (50 mg/kg group), or day 9 (70 mg/kg group). Gross necropsy examination of all mice had no significant findings. No hematological or liver enzyme abnormalities were noted in any of the groups of mice. Histological examination of all tissues examined was essentially normal in all treatment groups.

volume of ≈ 60 mm^3, mice are injected with 30 mg/kg single i.v. bolus dose of md-LErafAON via tail vein. Approximately 0.75–1mL blood sample is obtained from each animal via cardiac puncture under methoxyflourane anesthesia in microtainers containing sodium heparin as an anti-coagulant at pre-dose and at 5 min, 15 min, 30 min, 1 h, 2 h, 4 h, 8 h, 24 h, and 48 h after md-LErafAON administration (samples are collected from two mice per time point). Mice are euthanized and liver, spleen, kidneys, lungs, heart, and PC-3 tumor tissues are rapidly excised, rinsed in ice-cold normal saline, and snap frozen on dry ice. Blood samples are centrifuged at 3000 rpm for 10 min at 4°C to separate the plasma. The plasma and tissue samples are stored at −80°C until further analysis.

3.3.3. Isolation and Quantification of Antisense raf Oligodeoxyribonucleotide (rafAON)

The rafAON is isolated from plasma samples using phenol–chloroform extraction method and from tissue samples using a DNA extraction kit (Stragene, La Jolla, CA, USA). The rafAON concentration standards are prepared by adding known amounts of rafAON into blank plasma, followed by phenol–chloroform extraction (*see* **Note 9**). The extracts are loaded onto 20% polyacrylamide/ 8M urea gels and electrophoresed in TBE buffer. The gels are electroblotted onto nylon plus membrane in 0.5 × TBE buffer at 20 V for 1 h, and the blots are probed with γ^{32}P-ATP-labeled sense rafAON (rafSON) in ExpressHyb Hybridization Solution (Clontech, Palo Alto, CA, USA) at 35°C for 2 h. The radiolabeled probe is generated by the 5′-end labeling of rafSON with γ-^{32}P-ATP using T4 polynucleotide kinase and purification over Chroma Spin-10 columns (Clontech, Palo Alto, CA, USA) (*see* **Notes 10 and 11**). The autoradiographs are scanned using a computer program (Image-Quant software, Molecular Dynamics, Sunnyvale, CA, USA) and the amounts of AON in various samples are calculated by comparison to standards (*see* **Notes 12 and 13**).

3.3.4. Data Analysis

The elimination rate constant (β) is calculated from the linear regression analysis of plasma concentration–time curve. The area under the curve (AUC$_{0\rightarrow\infty}$) is calculated using the linear trapezoidal method with extrapolation of the terminal phase to infinity (C_{last}/β), where C_{last} is the last measured concentration. Other parameters calculated are as follows: Total body clearance (Cl) = dose/AUC; volume of distribution (V_{area}) = Cl/β; and elimination half-life ($t_{1/2\beta}$) = 0.693/β (*see* **Note 14**).

Examples of the gel electrophoresis method to establish the rafAON concentration standards prepared in control human plasma and a representative standard curve are shown in **Fig. 3a**. Example of validating the integrity of rafAON (15-mer) is shown in **Fig. 3b**. Examples of the pharmacokinetics and tissue distribution profiles of md-LErafAON in PC-3 tumor-bearing mice are shown in **Figs. 4 and 5**, and **Tables 1 and 2** (see **Notes 15 and 16**).

3.4. Therapeutic Efficacy of a Combination of Systemically Delivered md-LErafAON and Ionizing Radiation (IR)

3.4.1. Tumor Xenografts in Mice, Treatments, and Monitoring of Tumor Growth

Logarithmically growing PC-3 cells are injected s.c. (4×10^6 cells in 150µL of phosphate buffered saline, PBS) into the left flank region of male Balb/c nu/nu

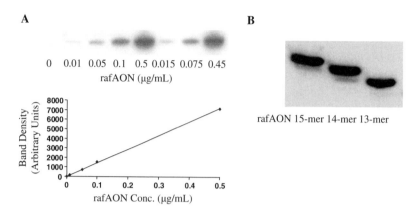

Fig. 3. (**a**) Sensitivity of detection of the standard rafAON (15-mer) in plasma. Control human plasma samples were spiked with indicated concentrations of rafAON, and rafAON was extracted by phenol–chloroform, resolved by 20% polyacrylamide/8*M* urea gel, followed by probing with ^{32}P-labeled complementary oligo, rafSON. Autoradiographs were scanned using a computer program (Image-Quant software). Similar experiments were performed with LErafAON and no lipid interference was detected in the extraction or quantification process. ***Top panel***: Autoradiographic representation of rafAON concentration standards detected by the gel electrophoresis method. ***Bottom panel***: Representative rafAON concentration standard curve plotted using data obtained from one blank (0 μg) and four standard samples (0.01, 0.05, 0.1, and 0.5 μg/mL). (**b**) Integrity of rafAON (15-mer). Custom-synthesized (Gibco BRL Life Technologies) rafAON size markers (15-mer: GTG CTC CAT TGA TGC; 14-mer: GTG CTC CAT TGA TG; 13-mer: GTG CTC CAT TGA T) were stored as lyophilized, as recommended by the manufacturer (the underlined bases are phosphorothioated). These markers are used as size markers to identify potential degradation products of rafAON in experimental samples. The markers were reconstituted in normal saline prior to each use. The rafAONs of various sizes were resolved by 20% polyacrylamide/8*M* urea gel (approximately 75–100 ng per lane) and probed with ^{32}P-labeled complementary sense raf Oligo.

mice. Tumors are allowed to grow to a mean tumor volume of ~60 mm[3] before initiation of the treatment. Tumor volumes are determined from caliper measurements of the three major axes () and calculated using *abc*/2, an approximation for the volume of an ellipse (Π *abc*/6). Tumor-bearing mice are randomly divided into various treatment groups (*n* = 6–8). The first day of treatment is designated as day 0. md-LErafAON and md-LErafAON+IR treatment groups receive 10 i.v. injections of 25 mg/kg/dose md-LErafAON over 14 days (day 0, 1, 3, 4, 6, 7, 9, 10, 12, and 13). Radiation is delivered locally to the tumors of mice in the IR alone and md-LErafAON+IR treatment groups once daily on days 3, 4, 7, 8, 9, 10, and day 11 using a [^{137}Cs] irradiator (J.L. Shepard Mark I) (2.37 Gy/min, 2.1 Gy/day). In the combination group, the two treatments are given 6–8 h

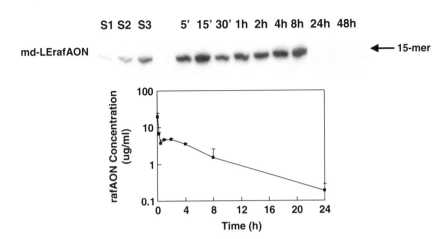

Fig. 4. The plasma concentration–time profile of md-LErafAON in male Balb/c nu/nu mice. PC-3 tumor cells (5×10^6 cells) were injected s.c. into the left flank region of male Balb/c nu/nu mice. Tumor xenografts were grown to a mean tumor volume of ≈ 60 mm^3. Mice were injected i.v. with a single dose of 30 mg/kg md-LErafAON, and rafAON concentrations in plasma samples collected at various times were determined by denaturing gel electrophoresis. *Top panel*: Representative autoradiograph showing intact rafAON in plasma at various time points post-dose. S1, S2, and S3 are rafAON standards prepared using blank plasma. *Bottom panel*: Plasma concentration–time curve of md-LErafAON. Following i.v. administration of 30 mg/kg md-LErafAON, the peak plasma concentration of 19.68 ± 5.89 µg/mL was achieved in Balb/c nu/nu mice bearing PC-3 tumor within 5 min after administration of md-LErafAON, and intact rafAON (15-mer) could be detected up to 24 h. Quantification data were calculated based on comparison with known concentrations of the standard samples, and then normalization against sample dilution factors used for loading. Each point represents mean $\pm$ SD ($n = 2$).

apart with md-LErafAON treatment being given first. Control groups receive i.v. doses of BL or md-LEMM on the same dosing and schedule as md-LErafAON or are left untreated (*see* **Notes 17–19**). Tumor sizes are monitored once or twice weekly for 32days post-treatment initiation. Individual tumor volume is calculated as the percentage of pre-treatment tumor volume (day 0, the first day of dosing; 100%), and the mean tumor volume (% initial) $\pm$ S.E. for each treatment group is plotted.

3.4.2. Detection and Quantification of Raf-1 Expression in Tumor and Normal Tissues

Normal and tumor tissues of PC-3 tumor-bearing Balb/c nu/nu mice are excised in various treatment groups within 6–12 h after the last treatment.

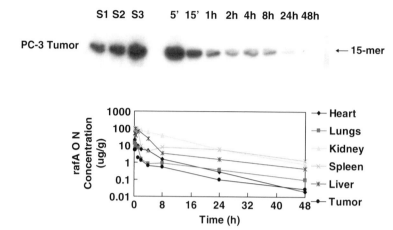

Fig. 5. The tissue concentration–time profile of md-LErafAON in PC-3 tumor-bearing male Balb/c nu/nu mice. PC-3 tumor cells (5×10^6 cells) were injected s.c. into the left flank region of male Balb/c nu/nu mice. Tumor xenografts were grown to a mean tumor volume of ≈ 60 mm^3. Mice were injected i.v. with a single dose of 30 mg/kg md-LErafAON. The rafAON concentration in various tissue samples was determined by denaturing gel electrophoresis. ***Top panel***: Representative autoradiograph showing intact rafAON in PC-3 tumor tissues excised at various time points post-dose. S1, S2, and S3 are rafAON standards prepared using pre-dose tumor tissue homogenate. ***Bottom panel***: Tissue concentration–time curve of md-LErafAON. Quantification data were calculated based on comparison with known concentrations of the standard samples, and then normalization against the weights of organs collected. Liver, spleen and kidneys were the preferential sites of rafAON accumulation with intact rafAON (15-mer) detectable up to at least 48 h. Remarkably, intact rafAON was also detected in PC-3 tumor tissue for up to at least 48 h. A peak rafAON concentration of 16.61 ± 11.90 μg/g tumor tissue was obtained at 30 min to 1 h post-dose ($n = 2$).

Table 1
Plasma Pharmacokinetic Parameters of md-LErafAON in PC-3 Tumor-Bearing Balb/c nu/nu Mice After Single i.v. Dose of 30 mg/kg

C_{max} (μg/mL)	$t_{1/2\alpha}$ (h)	$t_{1/2\beta}$ (h)	$AUC_{0 \to \infty}$ (μg.h/mL)	Cl (L/h/kg)	V_d (L/kg)
19.68 ± 5.89	0.06 ± 0.02	4.99 ± 0.89	47.13 ± 10.32	0.65 ± 0.14	4.61 ± 0.19

C_{max} = peak plasma concentration; $t_{1/2\alpha}$ = distribution half-life; $t_{1/2\beta}$ = elimination half-life; AUC = area under the plasma concentration–time curve; Cl = total body clearance; V_d = volume of distribution. Values shown are mean $\pm$ SD ($n = 2$).

Tissues are stored at –80°C until use. Raf-1 protein expression is examined in whole cell extracts from tissues by Western blotting with monoclonal anti-Raf-1 antibody (Transduction laboratories, Lexington, KY, USA) (1:17,000), followed by reprobing the same blot with polyclonal anti-G3PDH antibody (Trevigen,

Table 2
Tissue Distribution: $AUC_{0 \rightarrow 48h}(\mu g.h/g)$ of md-LErafAON in PC-3 Tumor-Bearing Balb/c nude Mice After Administration of Single i.v. Dose of 30 mg/kg

Heart	Liver	Spleen	Lungs	Kidneys	PC-3 tumor*
61.63	309.90	321.86	36.64	919.86	16.61 ± 11.90

*Value shown is mean $\pm$ SD ($n = 2$).

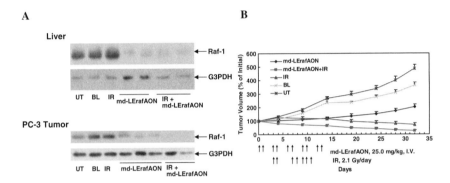

Fig. 6. Systemically administered md-LErafAON inhibits Raf-1 protein expression in normal and tumor tissues, and md-LErafAON treatment in combination with IR enhances prostate tumor growth regression. PC-3 tumor cells (5×10^6 cells) were injected s.c. into the left flank region of male Balb/c nu/nu mice. Tumor xenografts were grown to a mean tumor volume of ≈ 60 mm^3. Animals were randomized into five groups ($n = 6$–8). Day 0 represents the first day of treatment. Mice were treated with 25 mg/kg/dose i.v. md-LErafAON ($\times$ 10; day 0. 1, 3, 4, 6, 7, 9, 10, 12, 13) and/or 2.1 Gy/day of IR on day 3, 4, 7, 8, 9, 10 and 11. Control groups received either BL at the same dosing schedule as md-LErafAON or no treatment (UT). Tumor growth was monitored at least twice a week for up to day 32 post-treatment initiation. Representative animals in each treatment group were sacrificed within 6–12 h after the last treatment, and liver and tumor tissues were collected for Raf-1 expression analysis. (**a**) Inhibition of Raf-1 protein expression in liver and tumor tissues of mice treated with md-LErafAON. Within 24 h after the last dose of md-LErafAON, Raf-1 protein expression in liver and tumor tissues was approximately 9% and 29%, respectively, compared to UT control group (100%). There was no change in expression of B-Raf protein following md-LErafAON treatment (data not shown). Systemic treatment with liposome-entrapped mismatch oligonucleotide (md-LEMM, i.v., 25 mg/kg/dose, $\times$ 10) had no effect on Raf-1 or B-Raf expression in tumor tissues (data not shown). (**b**) The anti-tumor activities of md-LErafAON, IR, and combination treatments in PC-3 tumor-bearing athymic mice. The combination of md-LErafAON and IR treatments caused a steady decline in the mean %initial tumor volume for at least up to 19days post-last treatment.

Gaithersburg, MD, USA) (1:10,000). Raf-1 levels are quantified using Image-Quant software (Molecular Dynamics) (see **Notes 20–22**).

3.4.3. Statistical Analysis

One-way ANOVA is performed to determine the statistical significance of treatment-related changes in mean tumor volume in Balb/c nu/nu mice.

Examples of the effects of systemically delivered md-LErafAON on Raf-1 expression in normal and tumor tissues and on prostate tumor growth in athymic mice are shown in **Fig. 6**.

4. Notes

1. The terminal base linkage at the 5′ and 3′ ends of these sequences was modified to a phosphorothioate group.
2. The sequence of the good manufacturing practice (GMP) regulations grade product (97% pure) is confirmed by MALDI-TOF mass spectrometry by the manufacturer.
3. All other chemicals are of reagent grade.
4. Alternative method known as the thin-film method can be used to prepare md-LErafAON. Cationic liposomes are prepared using DMTAP, PC, and CHOL in a molar ratio of 1:3.2:1.6. The lipids (5 mg DMTAP, 20 mg PC, and 5 mg CHOL) are dissolved in 2mL of chloroform and evaporated to dryness in a round-bottomed flask using a rotatory vacuum evaporator in a 37°C water bath for approximately 20 min. Liposome-entrapped antisense *raf* oligodeoxyribonucleotide (md-LErafAON) is prepared by hydrating the dried lipid film overnight at 4°C with 1mL of rafAON at 2 mg/mL in normal saline. The film is dispersed by vigorous vortexing and the liposome suspension is sonicated for 10 min in a bath type sonicator (Model XL 2020, Misonix, Inc., Farmingdale, NY, USA). The rafAON to lipid ratio is 1:15 (w/w). The BL are prepared exactly as described above in the absence of rafAON. The liposome-encapsulated rafAON is stored at 4°C and used within 3days after preparation.
5. The hydration time of 30 min has been shown to work well *(17)*. The vial should be gently swirled a couple of times prior to subsequent use.
6. The liposome-entrapped rafAON is stored at 4°C and used within 3days after preparation. Entrapment efficiency is verified using at least three independent preparations, and each preparation tested in triplicate.
7. The entrapment efficiency is tested immediately after formulation (0 h post-preparation) and on day 8 post-preparation. The liposome-entrapment efficiency of rafAON is unchanged for up to at least 1 week after formulation.
8. Clinical pathology is conducted on each animal independently. The report from the consultant laboratory includes evaluation of total bilirubin, urea nitrogen, creatinine, alkaline phosphatase (ALP), aspartate aminotransferase (AST), alanine aminotransferase (ALT), WBC, RBC, and platelet count. It is important to note that the accuracy of most assays is affected by hemolysis.

9. *Preparation of the rafAON concentration standards.* The rafAON concentration standards are prepared by adding known amounts of rafAON into blank plasma, followed by extraction of rafAON using the phenol–chloroform extraction method. Spike known concentration of rafAON (50–250 ng) into control plasma to prepare standards. Take 100μL of standard. Extract rafAON with 25μL of phenol–chloroform. Centrifuge at 14,000 rpm for 10 min. Take the supernatant, thus the extracted sample. Prepare 20% polyacrylamide/8M urea gel. Take 15μL of extracted sample and mix with equal volume of loading buffer, boil the sample for 4 min and chill in wet ice for 4 min. Load onto gel. Run the gel at 450 V for 5.5 h in 1 × TBE buffer. Electroblot the gel onto nylon membrane in 0.5 × TBE buffer at 20 V for 1 h. Dry the nylon membrane in open air, followed by UV crosslinking (120 mJ). Pre-hybridize the nylon membrane at 35°C for 2 h by using ExpressHyb™ Hybridization Solution. Hybridize the membrane at 35°C for 2 h by using 10- to 50-fold excess of the probe in ExpressHyb™ Hybridization Solution. Wash the membrane twice in wash solution 1 (2 × SSC, 0.1% SDS) at room temperature for 15 min each. Wash the membrane twice in wash solution 2 (0.2 × SSC, 0.1% SDS) at room temperature for 15 min each. Expose the nylon membrane to X-ray film at –80°C. Adjust the time of exposure according to the density of the bands. Scan the autoradiograph using ImageQuant software (Amersham Pharmacia Biotech, Piscataway, NJ, USA). The amounts of rafAON (15-mer) loaded in various lanes are plotted against the densities of the corresponding bands (arbitrary value) and the lowest and highest limits of detection of the standard within a linear range, i.e., lower limit of quantification (LLOQ) and upper limit of quantification (ULOQ) are established (**Fig. 3a**). The intact rafAON is a 15-mer. The rafAON size markers (15-mer, 14-mer, and 13-mer) are prepared in normal saline and run on the same gel to clearly identify any degradation products in experimental samples. An example of the detection of various sizes of rafAON is shown in **Fig. 3b**.

10. *Radiolabeling of the probe.* Each 15μL of labeling reaction contains: 6μL raf-SON (600 ng), 6μL γ-^{32}P-ATP (60 μCi), 1.5μL T4 polynucleotide kinase buffer, and 1.5μL T4 polynucleotide kinase. Incubate at 37°C for 90 min. Add 35μL of Tris/EDTA (TE) buffer (10mM Tris–HCl, 1mM EDTA, pH 8) into the reaction tube to terminate the reaction. Remove the unincorporated radioisotope by microcentrifugation using the Chroma Spin-10 column. Place the Chroma Spin-10 column into one 2-mL microcentrifuge tube. Centrifuge at 700 × g for 5 min to purge the equilibration buffer from the column. Remove the spin column and place the spin column into a new 2-mL microcentrifuge tube. Apply 50μL of radiolabeled rafSON to the center of the gel bed's flat surface. Centrifuge at 700 × g for 5 min. The purified probe is at the bottom of the 2mL centrifuge tube. Determine cpm/μL of the probe.

11. A 10- to 50-fold excess of the probe is used to ensure saturation of all bands.

12. A wide range of the rafAON sizes (15-mer–11-mer) and at least three standard concentrations (0.01–0.5 μg/mL) of rafAON within the concentration range of linearity should be included during gel electrophoresis to verify integrity and generate validated quantification of rafAON in plasma and tissue homogenates.

The calibration standards representing LLOQ, ULOQ, and a midpoint serve as quality control samples. These standards are freshly prepared in control plasma, extracted, and run in parallel to the all gels.

13. *Quantification and data analysis.* Compare the signal with the rafAON standard marker (15-mer) to check single or multiple bands in the sample. Bands below 15-mer indicate degradation products. By using the known concentrations of the 15-mer standard and their corresponding reading, calculate the concentration in experimental sample. If the sample values exceed the ULOQ, the samples are reassayed after dilution in control plasma. If the values are below the LLOQ on the standard curve, the data are reported as not evaluable or below the limit of detection (BLOD).

14. Plasma pharmacokinetic parameters are assessed by standard methods *(11–13,29)*.

15. *Human specimens.* In our laboratory, we first developed a good laboratory practice (GLP)-validated procedure for quantification of intact rafAON in control human plasma. The rafAON assay validation endpoints were standard curve, between-run precision and accuracy, within-run precision and accuracy, effects of dilution and freeze thaw, stability of rafAON at –80°C, and 4°C in plasma for various times, specificity, integrity of rafAON during plasma sample collection and processing, and lipid interference. The reader is referred to a previous citation for further details *(17,27)*.

16. *Biohazard handling.* All sample handling and disposal procedures should be performed in compliance with the guidelines of the institutional Environmental Health and Safety Office. Gloves should be worn when handling the samples. Discard tips, pipettes, and tubes into a radioactive waste bag, or an autoclavable biohazard bag.

17. In addition to the untreated and BL-treated control groups, additional controls such as liposome-entrapped mismatch oligonucleotide and/or liposome-entrapped non-specific ASO should also be included to rule out the "off-target" effects.

18. Initially, pilot in vivo studies should be performed using liposomal formulations of multiple ASO sequences (15-mer) designed against different regions of the target mRNA sequence. The purpose here is not only to identify the most potent and gene-specific ASO sequence but also to standardize the in vivo dose and dosing frequency, which may vary depending on the half-life of the target of interest.

19. The cationic liposomal formulation described here can be used for systemic administration of a minimally modified (end base modification) or fully modified (all bases) oligonucleotide sequence.

20. Target inhibition in tumor and normal tissues should be analyzed within 24 h after the last treatment.

21. Secondary antibody (polyclonal, anti-mouse or anti-rabbit) dilution is 1:10,000.

22. The Western blot should be reprobed with at least one control antibody, e.g., B-Raf (polyclonal antibody, 1:500) (Santa Cruz) for verification of target specificity of md-LErafAON for Raf-1.

Acknowledgments

We thank Dr. Lisa Portnoy, Ms. Laura Rutter-Call, Ms. Carrie Silver, and Ms. Amy Durham for technical assistance, and our colleagues and collaborators for their comments. PC-3 cells were obtained from the Tissue Culture Shared Resource, and microscopy was performed at the Microscopy Shared Resource of the Lombardi Comprehensive Cancer Center, Georgetown University Medical Center. Mice clinical pathology and histopathology tests were performed by ANTECH Diagnostics, and data evaluated by a board-certified veterinary pathologist (Pathology Associates International). The liposomal formulation and its use presented in this study are patented (United States patent #s 6126965, 6333314, 6559129, and 7262173). The research work was supported by grants from the National Institutes of Health and NeoPharm, Inc.

References

1. Kasid, U., Pfeifer, A., Brennan, T., Beckett, M., Weichselbaum, R.R., Dritschilo, A., and Mark, G.E. (1989) Effect of antisense c-raf-1 on tumorigenicity and radiation sensitivity of a human squamous carcinoma. *Science* **243**, 1354–1356.
2. Soldatenkov, V.A., Dritschilo, A., Wang, F-H., Olah, Z., Anderson, W.B., and Kasid, U. (1997) Inhibition of Raf-1 protein kinase by antisense phosphorothioate oligodeoxyribonucleotide is associated with sensitization of human laryngeal squamous carcinoma cells to gamma radiation. *Can. J. Sci. Am.* **3**, 13–20.
3. Iorns, E., Lord, C.J., Turner, N., and Ashworth, A. (2007) Utilizing RNA interference to enhance cancer drug discovery. *Nat. Rev. Drug Discov.* **6**, 556–568.
4. Monia, B.P., Johnston, J.F., Geiger, T., Muller, M., and Fabbro, D. (1996) Antitumor activity of a phosphorothioate antisense oligodeoxynucleotide targeted against C-raf kinase. *Nat. Med.* **6**, 668–675.
5. Agrawal, S., Jiang, Z., Zhao, Q., Shaw, D., Cai, Q., Roskey, A., Channavajiala, L., Saxinger, C., and Zhang, R. (1997) Mixed-backbone oligonucleotides as second generation antisense oligonucleotides: In vitro and in vivo studies. *Proc. Natl. Acad. Sci. U.S.A.* **94**, 2620–2625.
6. Marshall, J.L., Eisenberg, S.G., Johnson, M.D., Hanfelt, J., Dorr, F.A., El-Ashry, D., Oberst, M., Fuxman, Y., Holmlund, J., and Malik, S. (2004) A phase II trial of ISIS 3521 in patients with metastatic colorectal cancer. *Clin. Colorectal Cancer* **4**, 268–274.
7. Soutschek, J., Akinc, A., Bramlage, B., Charisse, K., Constien, R., Donoghue, M., Elbashir, S., Geick, A., Hadwiger, P., Harborth, J., John, M., Kesavan, V., Lavine, G., Pandey, R.K., Racie, T., Rajeev, K.G., Röhl, I., Toudjarska, I.., Wang, G., Wuschko, S., Bumcrot, D., Koteliansky, V., Limmer, S., Manoharan, M., and Vornlocher, H.P. (2004) Therapeutic silencing of an endogenous gene by systemic administration of modified siRNAs. *Nature* **432**, 173–178.
8. Morrissey, D.V., Lockridge, J.A., Shaw, L., Blanchard, K., Jensen, K., Breen, W., Hartsough, K., Machemer, L., Radka, S., Jadhav, V., Vaish, N., Zinnen, S., Vargeese, C., Bowman, K., Shaffer, C.S., Jeffs, L.B., Judge, A., MacLachlan, I..,

and Polisky, B. (2005) Potent and persistent in vivo anti-HBV activity of chemically modified siRNAs. *Nat. Biotechnol.* **23**, 1002–1007.

9. Krützfeldt, J., Rajewsky, N., Braich, R., Rajeev, K.G., Tuschl, T., Manoharan, M., and Stoffel, M. (2005) Silencing of microRNAs in vivo with 'antagomirs'. *Nature* **438**, 685–689.

10. Baigude, H., McCarroll, J., Yang, C.S., Swain, P.M., and Rana, T.M. (2007) Design and creation of new nanomaterials for therapeutic RNAi. *ACS Chem. Biol.* **24**, 237–241.

11. Gokhale, P.C., Soldatenkov, V., Wang, F-H., Rahman, A., Dritschilo, A., and Kasid, U.(1997) Antisense raf oligodeoxyribonucleotide is protected by liposomal encapsulation and inhibits Raf-1 protein expression in vitro and in vivo: implications for gene therapy of radioresistant cancer. *Gene Ther.* **4**, 1289–1299.

12. Gokhale, P.C., McRae, D., Monia, B.P., Bagg, A., Rahman, A., Dritschilo, A., and Kasid, U. (1999) Antisense raf oligodeoxyribonucleotide is a radiosensitizer in vivo. *Antisense Nucleic Acid Drug Dev.* **9**, 191–201.

13. Gokhale, P.C., Zhang, C., Newsome, J., Pei, J., Ahmad, I., Rahman, A., Dritschilo, A., and Kasid, U. (2002) Pharmacokinetics, toxicity, and efficacy of ends-modified raf antisense oligodeoxyribonucleotide encapsulated in a novel cationic liposome (LErafAON). *Clin. Cancer Res.* **8**, 3611–3621.

14. Mewani, R.R., Tang, W., Rahman, A., Dritschilo, A., Ahmad, I., Kasid, U.N., and Gokhale, P.C. (2004) Enhanced therapeutic effects of doxorubicin and paclitaxel in combination with liposome-entrapped ends-modified raf antisense oligonucleotide against human prostate, lung and breast tumor models. *Int. J. Oncol.* **24**, 1181–1188.

15. Pei, J., Zhang, C., Gokhale, P.C., Rahman, A., Dritschilo, A., Ahmad, I., and Kasid, U. (2004) Combination with liposome-entrapped, ends-modified raf antisense oligonucleotide (LErafAON) improves the anti-tumor efficacies of cisplatin, epirubicin, mitoxantrone, docetaxel, and gemcitabine. *Anticancer Drugs* **15**, 243–253.

16. Pal, A., Ahmad, A., Khan, S., Sakabe, I., Zhang, C., Kasid, U., and Ahmad, I. (2005) Systemic delivery of RafsiRNA using cationic cardiolipin liposome silences Raf-1 expression and inhibits tumor growth in xenograft model of human prostate cancer. *Int. J. Oncol.* **26**, 1087–1091.

17. Zhang, C., Pei, J., Kumar, D., Sakabe, I., Boudreau, H.E., Gokhale, P.C., and Kasid, U.N. (2007) Antisense oligonucleotides: target validation and development of systemically delivered therapeutic nanoparticles. In: Target Discovery and Validation Reviews and Protocols, Vol. 2, Emerging Molecular Targets and Treatment Options (Sioud, M., ed). Humana Press, Inc. *Methods Mol. Biol.* **361**, pp 163–185.

18. Heidel, J.D., Yu, Z., Liu, J.Y., Rele, S.M., Liang, Y., Zeidan, R.K., Kornbrust, D.J., and Davis, M.E. (2007) Administration in non-human primates of escalating intravenous doses of targeted nanoparticles containing ribonucleotide reductase subunit M2 siRNA. *Proc. Natl. Acad. Sci. U.S.A.* **104**, 5715–5721.

19. Pirollo, K.F., Rait, A., Zhou, Q., Hwang, S.H., Dagata, J.A., Zon, G., Hogrefe, R.I., Palchik, G., and Chang, E.H. (2007) Materializing the potential of small interfering RNA via a tumor-targeting nanodelivery system. *Cancer Res.* **67**, 2938–2943.

20. Palliser, D., Chowdhury, D., Wang, Q.Y., Lee, S.J., Bronson, R.T., Knipe, D.M., and Lieberman, J. (2006) An siRNA-based microbicide protects mice from lethal herpes simplex virus 2 infection. *Nature* **439**, 89–94

21. Kasid, U. and Dritschilo, A. (2003) RAF antisense oligonucleotide as a tumor radiosensitizer. *Oncogene* **22**, 5876–5884.

22. Kasid, U., Suy, S., Dent, P., Ray, S., Whiteside, T.L., and Sturgill, T.W. (1996) Activation of Raf by ionizing radiation. *Nature* **382**, 813–816.

23. Suy, S., Anderson, W.B., Dent, P., Chang, E., and Kasid, U. (1997) Association of Grb2 with Sos and Ras with Raf-1 upon gamma irradiation of breast cancer cells. *Oncogene* **15**, 53–61.

24. Rudin, C.M., Holmlund, J., Fleming, G.F., Mani, S., Stadler, W.M., Schumm, P., Monia, B.P., Johnston, J.F., Geary, R., Yu, R.Z., Kwoh, T.J., Dorr, F.A., and Ratain, M.J. (2001) Phase I trial of ISIS 5132, an antisense oligonucleotide inhibitor of c-raf-1, administered by 24-hour weekly infusion to patients with advanced cancer. *Clin. Cancer Res.* **7**, 1214–1220.

25. Wilhelm, S., Carter, C., Lynch, M., Lowinger, T., Dumas, J., Smith, R.A., Schwartz, B., Simantov, R., and Kelley, S. (2006) Discovery and development of sorafenib: a multikinase inhibitor for treating cancer. *Nat. Rev. Drug Discov.* **5**, 835–844.

26. Rudin, C.M., Marshall, J.L., Huang, C.H., Kindler, H.L., Zhang, C., Kumar, D., Gokhale, P.C., Steinberg, J., Wanaski, S., Kasid, U.N., and Ratain, M.J. (2004) Delivery of a liposomal c-raf-1 antisense oligonucleotide by weekly bolus dosing in patients with advanced solid tumors: a phase I study. *Clin. Cancer Res.* **10**, 7244–7251.

27. Dritschilo, A., Huang, C.H., Rudin, C.M., Marshall, J., Collins, B., Dul, J.L., Zhang, C., Kumar, D., Gokhale, P.C., Ahmad, A., Ahmad, I., Sherman, J.W., and Kasid, U.N. (2006) Phase I study of liposome-encapsulated c-raf antisense oligodeoxyribonucleotide infusion in combination with radiation therapy in patients with advanced malignancies. *Clin. Cancer Res.* **12**, 1251–1259.

28. Boudreau, H.E., Broustas, C.G., Gokhale, P.C., Kumar, D., Mewani, R.R., Rone, J.D., Haddad, B.R., and Kasid, U. (2007) Expression of BRCC3, a novel cell cycle regulated molecule, is associated with increased phospho-ERK and cell proliferation. *Int. J. Mol. Med.* **19**, 29–39.

29. Gibaldi, M. and Perrier, D. (1982) Pharmacokinetics, 2nd edition. Marcel Dekker, New York, pp 45–111.

6

Peptide-Based Delivery of Steric-Block PNA Oligonucleotides

Saïd Abes, Gabriela D. Ivanova, Rachida Abes, Andrey A. Arzumanov, Donna Williams, David Owen, Bernard Lebleu, and Michael J. Gait

Summary

Several strategies based on synthetic oligonucleotides (ON) have been proposed to control gene expression. As for most biomolecules, however, delivery has remained a major roadblock for in vivo applications. Conjugation of steric-block neutral DNA mimics such as peptide nucleic acids (PNA) or phosphorodiamidate morpholino oligonucleotides (PMO) to cell penetrating peptides (CPP) has recently been proposed as a new delivery strategy. It is particularly suitable to interfere sequence-specifically with pre-mRNA splicing thus offering various applications in fundamental research and in therapeutics. The chemical synthesis of these CPP conjugates as well as methodologies to monitor their cellular uptake and their efficiency in a reliable and easy to implement assay of splicing correction will be described.

Key Words: Cell penetrating peptides; Peptide nucleic acids; Splicing correction.

1. Introduction

Synthetic oligonucleotides (ON) such as antisense ON, ribozymes, small interfering RNA (siRNA), microRNA (miRNA), triple-helix forming ON or aptamers have been widely used to control gene expression through specific interactions with RNA, DNA or even proteins. Numerous modifications have been proposed to improve the pharmacological properties of synthetic ON, for example to improve their metabolic stability or their affinity, or to increase their selectivity in target recognition. Moreover, the targeted macromolecule is in

From: *Methods in Molecular Biology, vol. 480: Macromolecular Drug Delivery*, Edited by: M. Belting
DOI 10.1007/978-1-59745-429-2_6, © Humana Press, a part of Springer Science+Business Media, LLC 2009

most cases intracellular. Further, free ON are not taken up efficiently by most cell types unless associated with nucleic acid-delivery vectors. Several transfection protocols (such as for instance electroporation or lipofection) are easily implemented with cultured established cell lines. Unfortunately, most transfection methods are unsuitable for in vivo use. Thus, delivery is still considered as a major roadblock for clinical applications of synthetic ON *(1,2)*.

The comparative evaluation of ON analogues and ON-delivery vectors in a reliable and easy to implement assay is now possible with the splicing correction assay proposed by Kole et al. *(3,4)*. Intronic point mutations in a β-thalassemic globin gene activate cryptic splice sites leading to the aberrant splicing of this intron and as a consequence to a non-functional protein. Masking of the mutated site with a steric-block ON re-orients the splicing machinery toward complete removal of the intron and leads to the production of a correctly spliced mRNA. This mutated intron has been introduced into the coding region of a reporter luciferase gene and the construction has been stably transfected in HeLa cells (**Fig. 1**).

This splicing correction assay has now been widely adopted, since it has a low background and provides a positive read-out with a large dynamic response. Neutral steric-block ON, such as peptide nucleic acids (PNA) *(5)* or phosphorodiamidate morpholino oligonucleotides (PMO) *(6)*, are particularly suitable for this purpose, since they cannot recruit RNase H, they hybridize with high affinity and selectivity to complementary RNA, and they are metabolically very stable. However, they cannot be transfected with most commercially available delivery vectors.

New strategies for the delivery of biomolecules have emerged with the discovery of cell penetrating peptides (CPP), which are small, generally basic amino acid-rich peptides that are internalized within most cell types. More importantly, they allow the cellular uptake of chemically conjugated biomolecules of various types, including ON, peptides or proteins *(7,8)*.

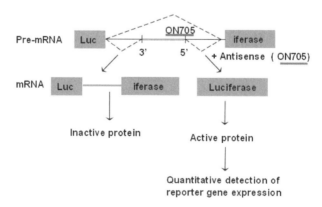

Fig. 1. Outline of the luciferase splicing correction assay.

PNA are synthesized by solid-phase Fmoc/Bhoc chemistry *(9–11)*. One to three Lys residues are generally added at the C-terminus to enhance aqueous solubility. After assembly (C to N), one additional Lys residue is added, followed by either a Cys residue (for disulfide conjugation) or by bromoacetylation (for thioether conjugation). Peptides are also synthesized by solid-phase Fmoc chemistry as C-terminal amides, which may help to enhance bio-stability, and with a Cys residue at the C-terminus for conjugation by either disulfide or thioether methods. It is possible to place a Cys residue anywhere in the peptide sequence provided conjugation be desired at other sites.

It is important to emphasize that a majority of human genes undergo alternative splicing and that several acquired (e.g., cancers, viral infections) or genetic diseases (e.g., β-thalassemia, Duchenne muscular dystrophy) can potentially be treated through ON-based control of splicing *(12)*. Once again, the efficient nuclear delivery of the correcting ON remains a major issue, and this is addressed by CPP-mediated delivery. Conjugation chemistries and assays to monitor the cellular uptake and the biological activity in a splicing correction assay of these PNA conjugates will be described.

2. Materials

2.1. Cell Culture and Cell Dissociation

1. HeLa cell cultures are propagated in Dulbecco's Modified Eagle Medium (D-MEM) (500 mL) (Gibco) supplemented with 10% fetal bovine serum (FBS, BioWest), 5 mL MEM non-essential amino acids (100 ×) (Gibco), 5 mL sodium pyruvate MEM (100 mM, Gibco), and 5 mL Penicillin–Streptomycin–Neomycin (PSN) Antibiotic Mixture (Gibco).
2. Mycoplasma Detection Kit from Roche Applied Science for routine screening of eventual mycoplasma contaminations.
3. Opti-MEM® I Reduced Serum Medium (1 ×) with L-Glutamine (Gibco) for serum-free experiments.
4. Trypsin–ethylenediamine tetraacetic acid (EDTA): 0.05% trypsin with 0.35 mM EDTA·4Na 1 × (Gibco) or pronase powder (Sigma–Aldrich).
5. Dulbecco's Phosphate Buffered Saline (D-PBS) (1 ×) (Gibco) for cell washing.
6. Forma Direct Heat CO_2 Incubator HEPA Class 100 (Thermo Electron Corporation) for cell cultures.
7. Laboratory laminar airflow cabinet BH-EN 2004-S. Type II, Catégorie 2 (Microbiological Safety Cabinets) for cell manipulations in sterile conditions.
8. Allegra™ 25R low-speed Beckmann centrifuge or Eppendorf Centrifuge 5417R for cell recovery.
9. Axiovert 40 C (transmitted light) (Carl Zeiss, Oberkochen, Germany) and Thoma cell for cell integrity routine checking and counting.

2.2. Synthesis of PNA–Peptide Conjugates

1. PNA synthesis reagents and materials: Use Fmoc-PAL-PEG-PS amide support (Applied Biosystems), and Fmoc (Bhoc) PNA monomers (Panagene or Link Technologies) and Fmoc(Boc)Lys from Novobiochem. Final Cys coupling is with Boc-Cys(Npys)-OH (Bachem) for Cys-terminated PNA or bromoacetic anhydride (made with bromoacetic acid and diisopropylcarbodiimide, both from Aldrich) for *N*-bromoacetyl PNA. Other synthesis reagents are piperidine (Romil, >99.5%), *N,N*-diisopropylethylamine (DIPEA, 99+%, Applied Biosystems), PyBop (Nova biochem), *N*-methylpyrrolidinone (NMP, ≥99.5% Aldrich), 2,6-lutidine (≥99%, Aldrich), acetic anhydride (Aldrich), and dimethylformamide (DMF, Merck/BDH). For final deprotection, use trifluoroacetic acid (TFA) obtained from Romil (>99.9%) and triisopropylsilane (TIS) from Aldrich (>99%). Use a polyethylene syringe (IST Empty Reservoir 1 mL, Kinesis) fitted with a 10-μm polyethylene frit and plastic tap (Kinesis). Carry out reversed-phase HPLC, for example using a Phenomenex Proteo C-18 column (analytical or semi-preparative) fitted with a heating jacket. Acetonitrile (Fisher Scientific, HPLC grade) and water (HPLC grade) are used as solvents.

2. Peptide synthesis reagents and materials: Use NovaSyn TGR resin (Novabiochem) for C-terminal amide synthesis and Fmoc amino acid monomers (Novabiochem) including Fmoc-Arg(Pbf)-OH, Fmoc-Asn(Trt)-OH, Fmoc-Cys(Trt)-OH, Fmoc-Gln(Trt)-OH, Fmoc-Glu(OtBu)-OH, Fmoc-His(Trt)-OH, Fmoc-Lys(Boc)-OH, and Fmoc-Trp(Boc)-OH. Other peptide synthesis reagents are as for PNA (above) with the addition of 1,2-ethanedithiol (EDT, >98%, Fluka). A Phenomenex Jupiter C-18 column (analytical and semi-preparative) may be used for reversed-phase HPLC.

3. Conjugation reagents: Formamide (>99.5%, Fluka) BisTris·HBr, ammonium acetate. HPLC columns are as for peptide and PNA syntheses and columns are immersed in a water bath (45°C) or surrounded by a very efficient heating jacket.

2.3. FACS Analysis of PNA and PNA–Peptide Conjugates Cellular Uptake and Cells Permeability

1. FACSCanto™ flow cytometer (BD Biosciences, San Jose, CA) using FACSDiva® software for PNA–peptide conjugates uptake.
2. WinMDI 2.8 free software to analyze the results.
3. Propidium iodide (PI, Molecular Probes, Eugene, OR) used at final concentration of 0.05 μg/mL for cell permeability quantification.

2.4. Fluorescence Microscopy Analysis of PNA–Peptide Intracellular Distribution

1. Alexa Fluor® 546-labeled Transferrin conjugate from Molecular Probes, Eugene, OR to stain endosomes.

2. Hoechst 33342 (trihydrochloride, trihydrate 10 mg/ml) from Molecular Probes, Eugene, OR, to stain nuclei (cell-permeant nuclear counterstain).
3. Zeiss Axiovert 200 M fluorescence microscope (Carl Zeiss, Oberkochen, Germany).
4. Adobe Photoshop CS2 software, ImageJ free software (NIH, USA, public domain), and Matrox Inspector software (Matrox Electronic System Ltd) for image treatment.

2.5. Luciferase Assay of Splicing Correction by PNA and PNA–Peptide Conjugates

1. Chloroquine from Sigma–Aldrich to increase endosome release of CPP–ON conjugates.
2. BCA^{TM}Protein Assay Kit (Pierce, Rockford, IL) and ELISA plate reader (Dynatech MR 5000, Dynatech Labs, Chantilly, VA) for the quantification of cellular protein concentrations.
3. Berthold Centro LB 960 luminometer (Berthold Technologies, Bad Wildbad, Germany) and Luciferase Assay System with Reporter Lysis Buffer from Promega for luciferase activity quantification.

2.6. RT-PCR Evaluation of Splicing Correction by PNA and PNA–Peptide Conjugates

1. Forward 5′-TTG ATA TGT GGA TTT CGA GTC GTC-3′ and reverse 5′-TGT CAA TCA GAG TGC TTT TGG CG-3′ luciferase primers from Eurogentec, Belgium.
2. RNA extraction using TRI REAGENT^{TM} from Sigma–Aldrich. Chloroform, isopropanol, and ethanol from Carlo Erba reagents.
3. Concentrator 5301 from Eppendorf for RNA pellets drying.
4. SuperScript III one-step RT-PCR system with Platinum$^®$ *Taq* polymerase (Invitrogen) and MJ Research PTC200 Peltier Thermal cycler for amplification.
5. Eppendorf BioPhotometer for amplification products' quantification.
6. Agarose and ethidium bromide powder from Sigma–Aldrich for gel electrophoresis using Amilabo electrophoresis power supply ST 1006T.
7. Lumi imager F1 Roche for image acquisition.
8. *Dpn*I, *Ava*I, and *Xba*I restriction enzymes for Promega pLuc705 plasmid DNA digestion.

3. Methods

3.1. Cell Culture and Cell Dissociation

1. Culture HeLa pLuc705 cells (may be purchased from Gene Tools,USA) as exponentially growing subconfluent monolayers in DMEM (Gibco) supplemented with

10% fetal calf serum (FCS), sodium pyruvate, non-essential amino acids, and antibiotics.

2. Wash cells twice with PBS and passage with trypsin–EDTA every other day in 175 cm² flasks for routine maintenance, for a maximum of 10 passages. For experiments, plate cells overnight on 24-well plates (1.75×10^5 cells/well).

3.2. Synthesis of PNA–Peptide Conjugates

3.2.1. Synthesis of N-Terminal Cys Functionalized and N-Bromoacetyl PNA

PNA with the desired N-terminal Cys or bromoacetyl functionalities may be purchased from Panagene (Korea). Alternatively, PNA synthesis may be achieved using an APEX 396 Robotic Peptide Synthesizer. Following final deprotection, the PNA is analyzed and purified using HPLC, and the molecular mass verified by MALDI-TOF mass spectrometry.

1. Dissolve each Fmoc-PNA monomer in NMP to give 0.2 *M* solutions (allow 200 μL per PNA or amino acid residue). Warming may be necessary in some cases.
2. Dissolve the PyBop in DMF to give a 0.2 *M* solution (allow 200 μL per PNA or amino acid residue).
3. Make a 0.2 *M* DIPEA, 0.2 *M* 2,6-lutidine solution in DMF (allow 200 μL per PNA or amino acid residue).
4. Make a 20% piperidine solution in DMF (allow 1.6 mL per PNA or amino acid residue).
5. Make a capping solution with 5% acetic anhydride and 6% 2,6-lutidine solution in DMF.
6. Weigh out 10 μmol solid support, put this into a reactor well and swell the support with 1 mL DMF for 15 min.
7. Programme the Synthesizer for the desired sequence (Lys residues are usually added to the C-terminus and N-terminus, to aid solubility, followed by the activated Cys monomer or bromacetic anhydride for the final coupling).
8. Start the synthesis. Each synthesis cycle consists of Fmoc-deprotection, double coupling, and capping (**Table 1**). Continue until the sequence is completed.
9. For Cys-terminated PNA, carry out the final coupling with Boc-Cys(Npys)-OH. Do not carry out a subsequent Fmoc deprotection step. For *N*-bromoacetyl PNA, carry out Fmoc deprotection followed by the final coupling as follows. Dissolve bromoacetic acid (2 mmol) in dichloromethane (5 mL) and add diisopropylcarbodiimide (2 mL, 0.5 *M*). Stir for 15 min and filter off the white precipitate (diisopropylurea) and evaporate the filtrate to approx. 2.5 mL. Adjust the volume to 6 mL with DMF and evaporate to approx. 4 mL by bubbling Argon gas through the solution. Filter the solution a second time. Use the resultant bromoacetic anhydride in DMF for coupling to the support.
10. Wash the support with DMF, then methanol and dry the support under vacuum in a desiccator.

Table 1
Synthesis Cycle for PNA or Peptide Synthesis on the APEX 396 Robotic Peptide Synthesizer

Synthesis step	Reagents and volumes (μL)	Time (min)
Deprotection	20% Piperidine soln. (800)	1
	20% Piperidine soln. (800)	4
Wash ($\times$ 5)	DMF (1000)	1
Couple ($\times$ 2)	PNA or amino acid (100)	30
	PyBop (100)	
	DiPEA/lutidine (100)	
	DMF (100)	
Wash ($\times$ 5)	DMF (1000)	1
Cap ($\times$ 2)	PNA capping soln. (1000)	5
Wash ($\times$ 5)	DMF (1000)	1

11. Place the support in a polyethylene syringe fitted with 10 μm frit and tap.
12. Simultaneously, cleave the PNA oligomer from the support and deprotect by adding 4 mL of a 95% TFA, 2.5% water, and 2.5% TIS for 4 h.
13. Filter into a 15-mL Falcon tube by washing the support with additional TFA (0.5 mL). Concentrate the filtrate to approx. 200 μL and precipitate the PNA oligomer with cold (4°C) diethyl ether.
14. Vortex the mixture and compact the precipitate by centrifugation (2500 rpm). Decant off the ether solution carefully and wash the precipitate with ether a further three times, compacting the residue and decanting off each time. *CAUTION: It is necessary to use sealed buckets when centrifuging solutions of flammable liquids such as ether.*
15. Analyze the crude product and purify by reversed-phase HPLC using an analytical or semi-preparative column, as appropriate, heated to 45°C. Monitor by UV at 260 nm. Buffer A: 0.1% TFA (aq.), Buffer B: 90% acetonitrile + 10% Buffer A (v/v).
16. A typical gradient for a 16–18 mer with 4 Lys residues is 5–20% Buffer B, which is used over 25 min.
17. Collect appropriate fractions, lyophilize, redissolve in water/acetonitrile as required, and analyze by HPLC and MALDI-TOF mass spectrometry.
18. Quantify the product by measuring the UV absorbance of an aliquot at 260 nm.

3.2.2. Synthesis of C-Terminal Cys-Containing Peptides

Peptides may be readily purchased from custom peptide synthesis suppliers. Alternatively, assemble the peptide (as a C-terminal amide) using a Peptide Synthesiser. The following protocol is designed for an APEX 396 Robotic Peptide Synthesiser.

1. Dissolve each amino acid derivative in DMF to give 0.5 M solutions (allow 600 µL per amino acid residue).
2. Dissolve the PyBop in DMF to give a 0.5 M solution (allow 600 µL per amino acid residue).
3. Make a 1 M DIPEA solution in DMF (allow 600 µL per amino acid residue).
4. Make a 20% piperidine solution in DMF (allow 2 mL per amino acid residue).
5. Weigh out 50 mg support (10 µmol), put into a reactor well and swell the support with 2 × 1 mL DMF (5 min each). Drain off the DMF from the well.
6. Programme the Synthesizer for the desired sequence and start the synthesis beginning with Fmoc deprotection and subsequent double couplings, but omitting the capping step (**Table 1**), and continue until the sequence is completed finishing with an Fmoc deprotection.
7. Wash the support with DMF, then propan-2-ol, and dry under vacuum in a desiccator.
8. Simultaneously, cleave the peptide from the support and deprotect with 94% TFA, 2.5% water, 2.5% EDT, and 1% TIS for 3–6 h.
9. Filter off the support and wash with additional TFA. Concentrate the filtrate to approx. 10% of the original volume and precipitate the peptide with cold (4°C) diethyl ether.
10. Vortex the mixture and compact the precipitate by centrifugation (2500 rpm). Decant off the ether solution carefully and wash the precipitate with ether a further five times, compacting the residue and decanting off each time. *CAUTION: It is necessary to use sealed buckets when centrifuging solutions of flammable liquids such as ether.*
11. Analyze the crude product and purify by use of reversed-phase HPLC using an analytical or semi-preparative column, as appropriate. Buffer A: 0.1% TFA (aq), Buffer B: 90% acetonitrile + 10% Buffer A (v/v).
12. Collect appropriate fractions, lyophilize, redissolve in water or Buffer A, and analyze by HPLC and MALDI-TOF mass spectrometry.

3.2.3. Synthesis of Disulfide-Linked Conjugates (Fig. 2)

1. Put into a microfuge tube 50 µL formamide (for lipophilic peptides add instead 25 µL formamide and 25 µL acetonitrile).
2. Add 20 nmol (2 µL of a 10 mM aqueous solution) of the (NPys)Cys-PNA oligonucleotide (from **Subheading 3.2.1.**) and then 50 nmol (5 µL of a 10 mM stock solution, 2.5 equivalents) of the Cys-functionalized peptide to be conjugated (from **Subheading 3.2.2.**).
3. Add 1 M NH$_4$Ac solution (10 µL).
4. Mix the solution by vortexing, centrifuge briefly and leave for 30–60 min at room temperature.
5. Purify the resulting conjugate by reversed-phase HPLC at 45°C using a single injection at flow rate of 1.5 mL/min. Use a gradient 15–35% B buffer over 25 min when conjugating to highly basic peptides or a gradient 10–60% B buffer when

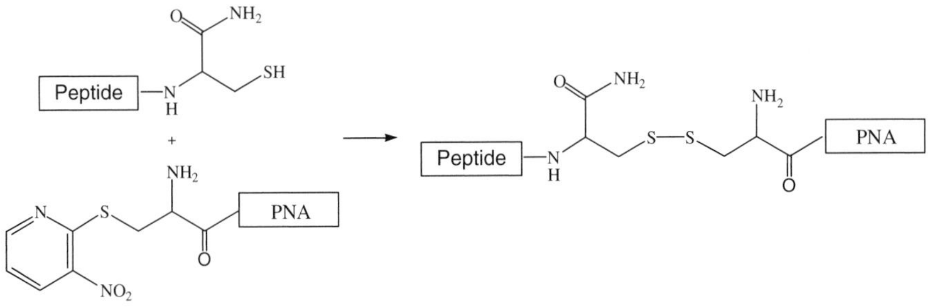

Fig. 2. Formation of a disulfide linkage.

conjugating to more lipophilic peptides. Buffers are the same as for peptides (*see* **Subheading 3.2.2.**).

6. Collect the product and lyophilize.
7. Dissolve the residue in sterile water, analyze by HPLC and by MALDI-TOF mass spectrometry, and quantify by measuring the absorbance at 260 nm.

3.2.4. Synthesis of Thioether-Linked Conjugates (**Fig. 3**)

1. Dissolve 50 nmol bromoacetyl PNA in 45 μL formamide and 10 μL BisTris·HBr buffer (pH 7.5).
2. Add 15.6 μL C-terminal Cys-containing peptide (8 m*M*, 125 nmol, 2.5 equivalents).
3. Heat the solution at 40°C for 2 h.
4. Purify the product by reversed-phase HPLC at 45°C. Gradients are similar to those in **Subheading 3.2.3.**
5. Collect the product and lyophilize.
6. Analyze by HPLC and by MALDI-TOF mass spectrometry and quantify by measuring the absorbance at 260 nm.

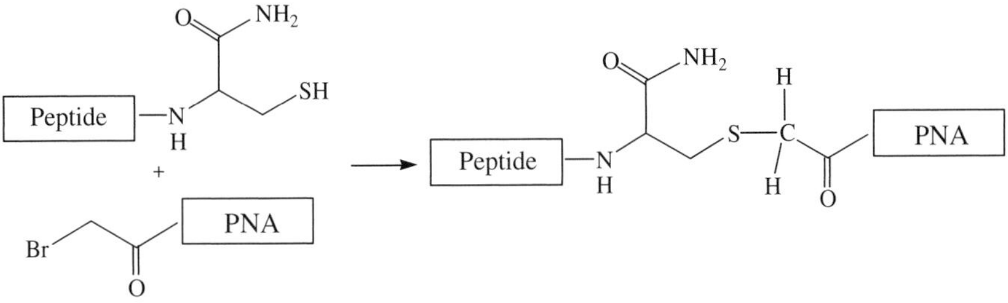

Fig. 3. Formation of a thioether linkage.

3.3. FACS Analysis of PNA–Peptide Conjugates Cellular Uptake and Cell Permeability Assay

1. Wash exponentially growing HeLa pLuc705 cells with PBS to remove cell culture medium, treat with trypsin–EDTA for 5 min, centrifuge at $900 \times g$ at 4°C for 5 min, wash twice with PBS, centrifuge again, resuspend in D-MEM, plate on 24-well plates (1.75×10^5 cells/well), and culture overnight.
2. Discard the culture medium and wash cells twice with PBS.
3. Discard PBS and incubate cells with fluorescently-labeled conjugates diluted in Opti-MEM or D-MEM.
4. After incubation for the appropriate period of time, wash the cells twice with PBS and treat with trypsin–EDTA (5 min, 37°C) or pronase (0.1%)/EDTA (1 mM) (5 min, 4°C).
5. Resuspend cells in PBS·5% FCS, centrifuge at $900 \times g$ (5 min, 4°C), and resuspend in PBS·0.5% FCS containing 0.05 μg/mL PI.
   Note: Use PI to analyze the effects of CPP–ON conjugates on cell permeability.
6. Analyze fluorescence with a FACS fluorescence activated sorter (BD Bioscience) for cellular uptake and PI permeabilization using WinMDI 2.8 free software. Exclude PI-stained cells from further analysis by appropriate gating. Analyze a minimum of 20,000 events per sample.

3.4. Fluorescence Microscopy Analysis of PNA–Peptide Intracellular Distribution

1. Detach exponentially growing HeLa pLuc705 cells (3.5×10^5) with trypsin (0.05%)–EDTA·4Na (0.35 mM), centrifuge at $900 \times g$ for 5 min, resuspend in 2 mL OptiMEM, and incubate with the fluorochrome-labeled conjugates (between 1 and 2.5 μM) for 30–120 min.
2. Treat the cells with trypsin and rinse twice with PBS.
3. Incubate the cells with Transferrin-Alexa 546 (10 μg/mL diluted in OptiMEM) for 10 min at 37°C to stain endosomes.
4. Wash twice with PBS.
5. Incubate with Hoechst (blue fluorescence) for 5 min to stain nuclei.
6. Wash twice with PBS and add 1 mL of complete medium.
7. Analyze the distribution of fluorescence in live unfixed cells on a Zeiss Axiovert 200 M fluorescence microscope with 63 × Plan-Apochromat objective, an Axio-Cam MRm camera, and Axiovision software.

3.5. Luciferase Assay of Splicing Correction by PNA and PNA–Peptide Conjugates

1. Detach exponentially growing HeLa pLuc705 cells with trypsin–EDTA, plate on 24-well plates (1.75×10^5 cells/well), and culture overnight.
2. Wash twice with PBS and incubate with the splice correcting conjugate or its scrambled version at the appropriate concentrations usually between 0.5 and 4 h in OptiMEM medium.

3. Wash cells and continue incubation for 20 h in complete D-MEM medium containing 10% FCS.
4. Wash cells twice with PBS and lyse with Reporter Lysis Buffer (Promega, Madison, WI).
5. Quantify luciferase activity in a Berthold Centro LB 960 luminometer (Berthold Technologies, Bad Wildbad, Germany) using the Luciferase Assay System substrate (Promega, Madison, WI). Perform all experiments in triplicate.
6. Measure cellular protein concentrations with the BCAProtein Assay Kit (Pierce, Rockford, IL) and read using an ELISA plate reader (Dynatech MR 5000, Dynatech Labs, Chantilly, VA) at 550 nm. Perform all experiments in triplicate.
7. Express luciferase activities as relative luminescence units (RLUs) per microgram of protein. Average each data point over the three replicates.

3.6. RT-PCR Evaluation of Splicing Correction by PNA and PNA–Peptide Conjugates

1. Extract total RNA using 1 mL of TRI REAGENT™/well after measurement of luciferase activity. Add 300 μL of chloroform, mix gently, and incubate for 10 min at room temperature.
2. Centrifuge at $12,000 \times g$ for 15 min at 4°C and add an equal volume of isopropanol to the aqueous phase. Mix gently and incubate for 10 min at room temperature.
3. Centrifuge at $12,000 \times g$ for 15 min at 4°C and resuspend the pellet in 1 mL of 75% ethanol. Mix and centrifuge at $12,000 \times g$ for 5 min at 4°C. Discard the supernatant. Evaporate off the ethanol using an Eppendorf Concentrator 5301 for 1 min at 60°C.
4. Add 50 μL of Nuclease Free Water.
5. Quantify total RNA using Eppendorf BioPhotometer and control quality by 1% agarose gel electrophoresis on Amilabo electrophoresis power supply ST 1006T.
6. Amplify total RNA using SuperScript III one-step RT-PCR system with Platinum® *Taq* polymerase in the presence of Luciferase specific primers with MJ Research PTC200 Peltier Thermal cycler.

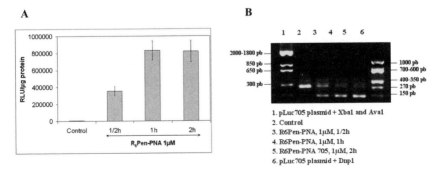

Fig. 4. Assay of splicing correction: Luciferase activity (**a**) and RT-PCR evaluation (**b**).

7. Analyse PCR products by electrophoresis using 2% agarose gel. Use digestion products of the plasmid pLuc705, digested by *Dpn*I, *Xba*I, and *Ava*I restriction enzymes, as molecular weight markers.

Splicing correction experiments by PNA-conjugates are shown in **Fig. 4**.

4. Notes

4.1. Cell Culture and Cell Dissociation

1. HeLa pLuc705 cells are stably transfected by a luciferase construction allowing the quantitative assessment of PNA–peptide conjugates nuclear delivery and biological activity (**Fig. 1**). Cells should not be passaged more than ten times and should be checked routinely (—one to two times per month) for the absence of mycoplasma contamination

4.2. Synthesis of PNA–Peptide Conjugates

2. There is no commercial Synthesiser currently recommended for PNA synthesis. We have found acceptable results using an APEX 396 Robotic Peptide Synthesiser and we have recently obtained good PNA assembly using a Liberty microwave Peptide Synthesiser. Key to success is minimization of times of piperidine treatment. Extended treatments can lead to a *trans*-acylation side reaction that will result in lower yields. PyBop must be dissolved freshly and on no account should be used after standing for more than 2 days.

3. In conjugation reactions, it is essential to maintain full solubility. Although not entirely essential in all cases, we prefer to maintain the presence of formamide in all conjugation reactions to ensure total solubility of both starting PNA and peptide and final conjugate. In the case of a very hydrophobic peptide (e.g., Transportan), the use of a mixture of formamide and acetonitrile may be helpful to maintain full solubility. Following conjugation, PNA–peptides can generally be purified by reversed-phase HPLC under acidic conditions, similar to the purification of both peptides and PNA. Some adjustment to the acetonitrile gradient conditions may be necessary from case to case. We recommend that conjugates are purified in one injection (i.e., as fast as possible after conjugation) and with use of a heated column (45°C) for optimal peak characteristics. Typically 60–80% conjugation yields should be achieved.

4.3. FACS Analysis of PNA-Peptide Conjugates Cellular Uptake and Cell Permeability Assay

4. For mechanistic studies, different drugs or treatments interfering with endocytosis may be used. In this case, pre-treat the cells with the inhibitors for the appropriate time and concentration. Inhibitors should also be present during incubation with the PNA–peptide conjugates. Most endocytosis inhibitors tend to be cytotoxic and should be used for the shortest possible period of time *(13)*.

5. Treatment with trypsin or pronase before FACS analysis is required to eliminate membrane-bound PNA–peptide conjugates *(14)*. Pronase is advantageous for some experiments (as for example when monitoring energy-dependence by low temperature incubation), since it is able to act at 4°C.

4.4. Fluorescence Microscopy Analysis of PNA–Peptide Intracellular Distribution

6. Experiments have to be performed on live cells, since most cell fixation protocols lead to artifactual re-distribution of PNA–peptide conjugates *(14)*.

4.5. Luciferase Assay of Splicing Correction by PNA and PNA–Peptide Conjugates

7. Co-treatment with 100 μM chloroquine may be included to improve endosomal release and to increase splicing correction. Chloroquine is required to achieve significant PNA or PMO nuclear delivery and splicing correction with Penetratin, $(Arg)_9$ or Tat_{48-60} at low concentrations. It is not required with recently described basic CPPs as R6Pen *(15)* or $(R-Ahx-R)_4$ *(16)*.
8. PNA–peptide conjugates should preferably be used at low concentrations (below 2.5 μM) to avoid cell permeabilization.

4.6. RT-PCR Evaluation of Splicing Correction by PNA and PNA–Peptide Conjugates

9. R6Pen *(15)* or $(R-Ahx-R)_4$ *(16)* PNA and PMO conjugates allow splicing correction in this assay with submicromolar EC_{50} values.
10. Programme used for reverse transcription and amplification:

    (a) Reverse Transcription: 1 cycle
    cDNA production: 30 min at 55°C
    Denaturation: 2 min at 94°C
    (b) Amplification: 30 cycles
    Denaturation: 20 s at 94°C
    Hybridization: 30 s at 60°C
    Elongation: 30 s at 68°C
    (c) Elongation: 1 cycle for 5 min at 68°C
    (d) Stock PCR products at –20°C.

Acknowledgements

We thank R. Kole (University North Carolina) for providing the pLuc705 cell line. Studies funded by EC grant QLK3-CT-2002-01989 and CEFIPRA grant 3205. S. Abes had a pre-doctoral fellowship from the Ligue Regionale contre le Cancer and R. Abes had a Region Languedoc-Roussillon training fellowship.

References

1. Torchilin, V. P. (2006) Recent approaches to intracellular delivery of drugs and DNA and organelle targeting. *Annu Rev Biomed Eng.* **8,** 343–375.
2. Zhang, X., and Godbey, W. T. (2006) Viral vectors for gene delivery in tissue engineering. *Adv Drug Deliv Rev.* **58,** 515–534.
3. Kang, S. H., Cho, M. J., and Kole, R. (1998) Up-regulation of luciferase gene expression with antisense oligonucleotides: implications and applications in functional assay development. *Biochemistry.* **37,** 6235–6239.
4. Kole, R., and Sazani, P. (2001) Antisense effects in the cell nucleus: modification of splicing. *Curr Opin Mol Ther.* **3,** 229–234.
5. Rasmussen, F. W., Bendifallah, N., Zachar, V., Shiraishi, T., Fink, T., Ebbesen, P., Nielsen, P. E., and Koppelhus, U. (2006) Evaluation of transfection protocols for unmodified and modified peptide nucleic acid (PNA) oligomers. *Oligonucleotides.* **16,** 43–57.
6. Summerton, J. (1999) Morpholino antisense oligomers: the case for an RNase H-independent structural type. *Biochim Biophys Acta.* **1489,** 141–158.
7. Abes, S., Richard, J. P., Thierry, A. R., Clair, P., and Lebleu, B. (2007) Tat-Derived Cell-Penetrating Peptides: Discovery, Mechanism of Cell Uptake, and Applications to the Delivery of Oligonucleotides. *Handbook of Cell-Penetrating Peptides (second edition).* 29–42.
8. Debart, F., Abes, S., Deglane, G., Moulton, H. M., Clair, P., Gait, M. J., Vasseur, J. J., and Lebleu, B. (2007) Chemical modifications to improve the cellular uptake of oligonucleotides. *Curr Top Med Chem.* **7,** 727–737.
9. Thomson, S. A., Josey, J. A., Cadilla, R., Gaul, M. D., Hassman, C. F., Luzzio, M. J., Pipe, A. J., Reed, K. L., Ricca, D. J., and Wiethe, R. W. et. al. (1995) Fmoc mediated synthesis of peptide nucleic acids. *Tetrahedron.* **51,** 6179–6194.
10. Braasch, D. A., Nulf, C. J., and Corey, D. A. (2002) Synthesis and purification of peptide nucleic acids. *Curr Protoc Nucleic Acid Chem.* 4.11.1–14.11.18.
11. Turner, J. J., Ivanova, G. D., Verbeure, B., Williams, D., Arzumanov, A. A., Abes, S., Lebleu, B., and Gait, M. J. (2005) Cell-penetrating peptide conjugates of peptide nucleic acids (PNA) as inhibitors of HIV-1 Tat-dependent trans-activation in cells. *Nucleic Acids Res.* **33,** 6837–6849.
12. Garcia-Blanco, M. A., Baraniak, A. P., and Lasda, E. L. (2004) Alternative splicing in disease and therapy. *Nat Biotechnol.* **22,** 535–546.
13. Richard, J. P., Melikov, K., Brooks, H., Prevot, P., Lebleu, B., and Chernomordik, L. V. (2005) Cellular uptake of unconjugated TAT peptide involves clathrin-dependent endocytosis and heparan sulfate receptors. *J Biol Chem.* **280,** 15300–15306.
14. Richard, J. P., Melikov, K., Vives, E., Ramos, C., Verbeure, B., Gait, M. J., Chernomordik, L. V., and Lebleu, B. (2003) Cell-penetrating peptides. A reevaluation of the mechanism of cellular uptake. *J Biol Chem.* **278,** 585–590.

15. Abes, S., Turner, J.J., Ivanova, G. D., Owen, D., Williams, D., Arzumanov, A., Clair, P., Gait, M. J. and Lebleu, B. (2007) Efficient splicing correction by PNA conjugation to an R6-Penetratin delivery peptide. *Nucleic Acids Res.* **35,** 4495–4502.

16. Abes, S., Moulton, H. M., Clair, P., Prevot, P., Youngblood, D. S., Wu, R. P., Iversen, P. L., and Lebleu, B. (2006) Vectorization of morpholino oligomers by the (R-Ahx-R)4 peptide allows efficient splicing correction in the absence of endosomolytic agents. *J Control Release.* **116,** 304–313.

7

Characterizing Peptide-Mediated DNA Internalization in Human Cancer Cells

Anders Wittrup and Mattias Belting

Summary

Cell penetrating peptides (CPPs) are currently used to deliver various macromolecular cargos to intracellular sites of action both in vitro and in vivo on an experimental basis. During the last few years, even more evidence has accumulated indicating that the main route of entry for most CPPs is through endocytosis rather than direct membrane penetration, as initially proposed. The specific endocytosis pathway utilized by CPPs is, however, still ill-defined and potentially varies depending on what CPPs, cargos, and cell lines are being studied. In this chapter, we provide detailed protocols for an initial characterization of the uptake mechanism involved in CPP-mediated delivery of DNA. Methods to both quantitatively and qualitatively study the uptake using fluorescence-assisted cell sorting (FACS) and confocal microscopy, respectively, are provided. Furthermore, methods to study the intracellular fate of the internalized cargo by co-localization studies between internalized DNA and established endosomal markers, e.g., transferrin, dextran as well as caveolin-1, are described. Finally, we provide a protocol to determine the dependence on dynamin, i.e., a central mediator of vesicle fission at the cell membrane, for DNA–peptide complex uptake using a dominant-negative construct of dynamin-2.

Key Words: HIV-tat; Cell penetrating peptide; Endocytosis; Proteoglycan; Gene delivery.

From: *Methods in Molecular Biology, vol. 480: Macromolecular Drug Delivery,* Edited by: M. Belting
DOI 10.1007/978-1-59745-429-2_7, © Humana Press, a part of Springer Science+Business Media, LLC 2009

1. Introduction

Protein transduction domains (PTDs) first identified in HIV-1 transactivator of transcription (TAT) *(1)* and in the homeobox transcription factor Antennapedia (with the PTD referred to as penetratin) *(2)* were shown to harbor the ability to deliver functionally active proteins and other macromolecules to the interior of the cells. As the family of the identified peptide sequences with such internalizing properties grew, the term cell penetrating peptides (CPPs) was introduced *(3)*, in reference to the proposed mode of action. However, during recent years an increasing number of reports have demonstrated that the entry of most CPPs is through a temperature-sensitive endocytotic process. Previous reports on temperature-independent plasma membrane penetration by CPPs were later ascribed to CPPs remaining on the cell surface due to insufficient washing or enzymatic ligand removal *(4)*. While there is relatively broad consensus on the notion that most CPPs enter cells through endocytosis, there are highly divergent views on which specific endocytotic pathway/s are involved. The HIV-TAT-derived PTD has, for example, been suggested to enter cells through clathrin-mediated endocytosis *(5)*, caveolar endocytosis *(6)* as well as macropinocytosis *(7)*. Further, it has been shown that the size of the internalized particles determines the route of entry *(8)*. The involvement of several different endocytotic pathways utilized by CPPs could thus, at least partly, be attributed to the differences between associated cargos and thereby to the size of the particles being studied. The vast majority of hitherto identified CPPs are highly cationic and interact with negatively charged cell surface proteoglycans, which mediate their internalization *(5,9,10)*. The confusion regarding which endocytotic pathway CPPs might utilize for macromolecular delivery also reflects the generally limited knowledge of the mechanism of proteoglycan-dependent internalization. Indeed, proteoglycan-dependent membrane transport [reviewed in ref. *(11)*] is thought to be responsible for the cellular uptake of virtually any non-viral transfection agent [e.g., polylysine *(12)*] as well as internalization of several viruses including herpes simplex *(13)* and HIV-1 *(14)*.

In this chapter, we provide protocols to determine the ability of a peptide to mediate DNA internalization in cultured human tumor cells. Fluorescence-assisted cell sorting (FACS) analysis is used to obtain quantitative data on the time and temperature dependence of macromolecular delivery. Confocal microscopy is used to study the subcellular localization in both fixed and live cells. Fluorescently labeled transferrin and dextran are used to label the clathrin-dependent *(15)* and the non-clathrin, non-caveolar *(16)* endocytic compartments, respectively. Expression of a caveolin-1-YFP fusion protein is used to label cell surface caveolae and intracellular caveosomes *(17)*. Finally a protocol, for the overexpression of dominant-negative dynamin [GTPase deficient dynamin-2 containing the amino acid substitution K44A *(18)*] is provided to evaluate the dynamin dependence of the uptake mechanism.

2. Materials

2.1. FACS Analysis of Peptide-Mediated DNA Uptake

1. Dulbecco's modified Eagle medium (DMEM, Invitrogen) with and without 10% fetal bovine serum (FBS, Sigma).
2. HeLa cells (ATCC) grown in DMEM with 10% FBS supplemented with glutamine and penicillum/streptomycin (Invitrogen).
3. Forty-eight-well cell culture plates (Nunc).
4. Flow cytometry tubes (BD).
5. Phosphate buffered saline (PBS, Invitrogen): 137 mM NaCl, 2.7 mM KCl, 10 mM Na$_2$HPO$_4$, 1.8 mM KH$_2$PO$_4$, pH 7.4.
6. PBS containing 1% bovine serum albumin (BSA, Sigma).
7. Trypsin/EDTA solution (Invitrogen).
8. DNA labeled with ULYSIS Alexa Fluor 488, using a Nucleic Acid Labeling Kit (Invitrogen). DNA provided in the kit or oligonucleotides from other suppliers (e.g., Sigma) can be used.
9. Peptide or protein of interest to study DNA transporting properties. As a positive control, we recommend using HIV-1 TAT-derived PTD (HIV-tat, GRKKRRQR-RRPPQC; corresponding to amino acids 48–60 of the HIV-1 TAT full length protein, *see* **Note 1**).
10. Flow cytometer equipped with a 488-nm argon laser and FITC and PI filters, e.g., FACScalibur or FACSaria (BD Biosciences).

2.2. Confocal Microscopy Analysis of Peptide-Mediated DNA Uptake

1. DMEM with and without 10% FBS.
2. HeLa cells grown in DMEM with 10% FBS supplemented with glutamine and-penicillin/streptomycin (Invitrogen).
3. Trypsin/EDTA solution.
4. DNA labeled with ULYSIS Alexa Fluor 488, using a Nucleic Acid Labeling Kit. DNA provided in the kit or oligonucleotides from other suppliers can be used.
5. Peptide or protein of interest to study DNA transporting properties, and HIV-tat peptide as positive control.
6. LabTek 8-well chamber slide (Nunc, *see* **Note 2**).
7. PBS.
8. PBS supplemented with 0.1% Triton X-100 (Sigma).
9. PBS supplemented with 1M NaCl.
10. PBS supplemented with 2% BSA.
11. Paraformaldehyde solution, 4%. Prepare a 50-mL solution by adding three drops of 1M NaOH to 40 mL water in a 50-mL tube. Dissolve 2 g paraformaldehyde (Sigma) by heating the tube to 56°C in a hot water bath. When the paraformaldehyde powder has dissolved completely, add 5 mL of 10× PBS and adjust pH to 7.4. Finally add water to bring the volume to 50 mL and filter the solution through a 0.2-μm sterile filter to remove any remaining particulate matter. The solution should be used immediately, or stored at –20°C.

12. Cover glass, 60×24 mm (Menzel-Gläser).
13. Permafluor mounting medium (Beckman Coulter).
14. Alexa Fluor 543-conjugated Phalloidin actin stain, diluted in methanol to 200 U/mL according to the instructions given by the manufacturer (Invitrogen).
15. TO-PRO-3 nuclear stain in 1 mM dimethyl sulfoxide (DMSO, Invitrogen).
16. Nail polish.
17. Immersion oil (Leica).
18. Confocal microscope system with argon-laser line at 488 nm and helium–neon-laser lines at 543 and 633 nm and appropriate filter configurations, such as the Leica SP2 or Zeiss Pascal Axioplan LSM.

2.3. Co-localization Studies of Internalized DNA and Endosomal Markers

1. All materials listed under **Subheading 2.2.**
2. DNA labeled with ULYSIS Alexa Fluor 647, using a Nucleic Acid Labeling Kit. DNA provided in the kit or oligonucleotides from other suppliers can be used.
3. Alexa Fluor 647-conjugated transferrin (Invitrogen).
4. Alexa Fluor 647-conjugated dextran (lysine fixable, MW 10 kDa; Invitrogen).
5. Electroporation cuvette, 4 mm (Sigma).
6. Mouse caveolin-1, C-terminal fused to yellow fluorescent protein (YFP; pEX_EF1_Cav-1-YFP; ATCC). The plasmid is supplied in an *Escherichia coli* clone and can be propagated using standard bacterial propagation techniques with kanamycin as selection pressure. Purify the plasmid using a maxi-prep kit from Qiagen, Sigma or other vendors.
7. Electroporator, such as the BTX ECM399 (Genetronics).

2.4. Disruption of Dynamin-Dependent Endocytosis Using Dominant-Negative Dynamin

1. All materials listed under **Subheading 2.2.**
2. Alexa Fluor 647-conjugated transferrin.
3. Electroporation cuvette, 4 mm.
4. Wild-type (WT) and dominant negative (K44A) Rat dynamin-2, N-terminal fused to Hemagglutinin (HA) (pcDNA3.1(-)HA-Dyn2 WT/K44A; ATCC). The plasmid is supplied in an *E. coli* clone and can be propagated using standard bacterial propagation techniques with ampicillin as selection pressure. Purify the plasmid using a maxi-prep kit from Qiagen, Sigma or other vendors.
5. Anti-HA mouse monoclonal antibody (12CA5, Roche).
6. Alexa Fluor 488-conjugated goat anti-mouse antibody (Invitrogen).
7. Electroporator, such as the BTX ECM399 (Genetronics).

3. Methods

Analysis of the uptake of fluorescent ligands is usually performed using either flow cytometry or fluorescence microscopy techniques. The advantage of the flow cytometry approach is its ability to measure the fluorescence of several thousands of cells, thereby obtaining more reliable quantitative data. Qualitative studies of endocytic processes usually include experiments, where drugs are used to disrupt different components of diverse cellular uptake mechanisms. The most common mechanisms of action of the drugs used include actin filament disruption, membrane cholesterol depletion, inhibition of the Na^+/H^+ anti-port, and inhibition of endosomal acidification. Hypertonic shock (often using sucrose) is also widely used to interfere with endocytic processes. All these methods are, however, more or less non-specific and the results of such studies should be interpreted cautiously and mainly as a complement to other more specific methods. In our opinion, it is preferable to perform drug treatment studies using flow cytometry, as quantifications are more reliable compared to microscopy techniques. *See* **Note 3** for examples of endocytosis disruptive drugs that can be used.

Most qualitative endocytosis studies require advanced microscopy techniques. Co-localization studies usually compare the staining pattern of a known endocytic component or ligand to the pattern of the studied ligand. In the case of studies of CPPs, data interpretation is complicated by the fact that most fixation protocols (including methanol and paraformaldehyde fixation) produce severe fixation artifacts with misallocation of the CPP and its cargo (mainly incorrect nuclear localization). The localization of small cationic peptides, i.e., the CPPs, is most prone to fixation artifacts. However, misallocations of larger macromolecules (e.g., oligonucleotides and plasmids) can never be ruled out and the ligand distribution in fixed and live cells should always be compared. It is therefore advisable to use fluorescent fusion proteins (e.g., caveolin-1-YFP) or fluorescently labelled endosomal markers (e.g., transferrin and dextran) in co-localization studies of CPPs, which will allow analysis of live cells.

3.1. Preparation of Samples for FACS Analysis

1. Seed 50,000 cells/well in 48-well cell culture plates in DMEM with 10% fetal calf serum (FCS) 24 h prior to ligand incubation.
2. Mix Alexa Fluor 488 labeled DNA and HIV-tat peptide (*see* **Note 4**) in DMEM without serum, allow to pre-complex for 20 min at room temperature (RT).
3. Wash the cells once with DMEM.
4. Add peptide/DNA solutions at appropriate time points to the washed cells (*see* **Note 5**).
5. Incubate the cells at 37°C for the desired period of time.

6. Remove the plate from the incubation chamber and place on ice, allow cooling for 15 min.
7. Add ice-cold peptide/DNA solution to one well for the 0 min incubation control (*see* **Note 6**).
8. Incubate for 30 min on ice in the dark.
9. Remove all peptide/DNA solutions.
10. Wash twice with 250 μL ice-cold PBS.
11. Add 150 μL trypsin solution to each well until the cells detach.
12. Add 1 mL ice-cold PBS containing 1% BSA to each well and transfer the suspension to labeled flow cytometry tubes.
13. Spin the samples at 400 × g, for 5 min, at 4°C.
14. Re-suspend the cell pellets in 1 mL ice-cold PBS containing 1% BSA and once again spin at 400 × g, for 5 min, at 4°C.
15. Re-suspend the resulting cell pellets in 250 μL ice-cold PBS containing 1% BSA. Keep the samples on ice until flow cytometry analysis.

3.2. Confocal Microscopy Analysis of Peptide-Mediated DNA Uptake

1. Seed approx. 20,000 HeLa cells/well in chamber slides in DMEM with 10% FCS, 24 h prior to incubation start.
2. Mix Alexa Fluor 488 labeled DNA and HIV-tat peptide (*see* **Note 4**) in DMEM without serum, allow to pre-complex for 20 min at RT, keep the solution at 4°C until use.
3. Wash the cells once with DMEM.
4. Take an aliquot of the peptide/DNA solution and warm to 37°C at each time point (*see* **Note 5**).
5. Add 150 μL of pre-warmed peptide/DNA solution to each well at the different time points.
6. Incubate the slide at 37°C for the desired period of time.
7. Remove the slide from the incubation chamber and place on ice, allow cooling for 15 min.
8. Add ice-cold peptide/DNA solution to one well for the 0 min incubation control.
9. Incubate for 30 min on ice in the dark.
10. Remove all peptide/DNA solutions.
11. Wash with 250 μL ice-cold PBS.
12. Wash once with 250 μL/well of 1 *M* NaCl in PBS to eliminate any remaining cell surface associated complex. Do not allow the cells to remain in the hyperosmotic buffer for more than 30 s (*see* **Note 7**).
13. Quickly remove the 1 *M* NaCl buffer and gently wash twice with PBS.
14. Remove the PBS. If the cells are to be fixed, proceed to **step 18**. If the cells are to be analyzed live continue with **step 15**.
15. Remove the plastic wells from the glass slide using the supplied tools (Nunc).
16. Carefully place a 60 × 24 mm cover glass on the slide and let the remaining PBS spread out by capillary force.

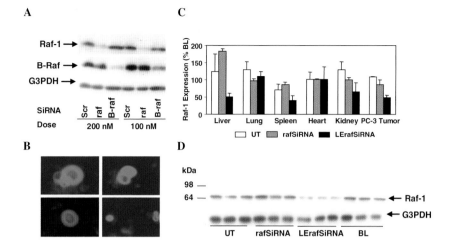

Color Plate 1. Effects of systemically delivered LErafsiRNA on Raf-1 expression in normal and tumor tissues. (*See* discussion and complete caption on p. 66.)

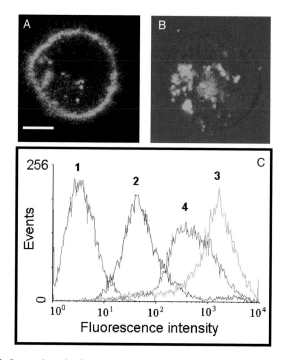

Color Plate 2. Low electric field-stimulated adsorption and uptake of BSA-FITC by COS 5–7 cells. Cell suspensions of COS 5–7 cells in DMEM-H supplemented with 6.8 μM BSA-FITC are subjected to LEF treatment of 20 V/cm (180 μs pulse duration at frequency of 500 Hz) for 1 min, followed by two successive washings with DMEM-H solution. Cells are observed by confocal microscopy. **a** and **b** are confocal images of FITC fluorescence taken at central optical section through the X–Y plane of COS 5–7 cells, in **b** it is superimposed also with phase contrast image (*blue channel*). Bar = 5 μm. **(a)** Imaging immediately after washing with DMEM-H. It can be seen that a large amount of BSA-FITC is adsorbed and some is already internalized. **(b)** Imaging after 25 min of incubation at 24 C followed by trypsinization. It resulted in an effective uptake of BSA-FITC and its removal from the plasma membrane. **(c)** Distribution histograms of BSA-FITC fluorescence intensity in COS 5–7 cells measured by flow cytometry 1 h after exposure to LEF and incubation at 24°C: *(1)* control cells in the absence of BSA-FITC in the medium; *(2)* control cells in the presence of BSA-FITC – constitutive uptake; *(3)* cells exposed to LEF in the presence of BSA-FITC – enhanced adsorption and uptake; *(4)* same exposed cells as in (3), but further subjected to trypsinization before measurement – enhanced uptake only (*See* discussion on p. 145.)

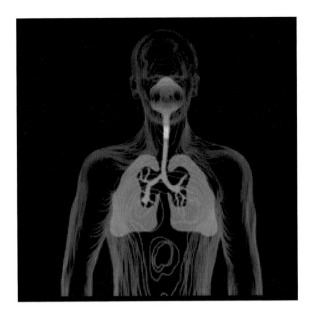

Color Plate 3. Particle deposition in the lungs. Particles larger than 10 μm are deposited in the mouth and throat (*yellow/orange area*) and are swallowed. Particles between 6 and 10 μm are deposited in the upper airways (*blue area*). Particles between 0.5 and 6 μm are within the respirable range and are deposited in the alveolar region (*pink area*). Image from: http://www.filterair.info/articles/article.cfm/ArticleID/ 36856F0C-747B-4E08-B730798D614269E9/Page/1 (*See* discussion on p. 166.)

17. Immediately, analyze the slide using a confocal microscope system equipped with an argon-laser line at 488 nm and appropriate filter configurations, such as the Leica SP2 or Zeiss Pascal Axioplan LSM.
18. Add 200 μL of freshly prepared or thawed 4% paraformaldehyde, incubate for 15 min at RT.
19. Wash twice with PBS.
20. Add 200 μL PBS with 0.1% Triton X-100 to permeabilize the cell membrane, incubate for 10 min at RT.
21. Wash three times with PBS.
22. Block the samples by adding 200 μL PBS with 2% BSA and incubate for 30 min.
23. Remove blocking solution and add 100 μL of PBS+2% BSA with TO-PRO-3 (1:100) and phalloidin-Alexa Fluor 543 (1:100) to label cell nuclei and actin filaments, respectively.
24. Wash three times with PBS, remove the PBS.
25. Remove the plastic wells from the glass slide using the supplied tools (Nunc).
26. Air-dry the slide by waving it in the air.
27. When the slide is almost dry, add a small drop of permafluor on two or more different parts of the slide to optimize fluorophor stability.
28. Carefully place a 60 × 24 mm cover glass on the slide and let the permafluor solution spread out by capillary force.
29. Keep the slide in the fridge overnight to allow the permafluor to solidify. To avoid the slide from drying out, seal the slit between the cover glass and the slide by applying a thin layer of nail polish over the slit. Avoid nail polish on the surface above or below the samples (*see* **Note 8**).
30. Analyze the slide using a confocal microscope system equipped with an argon-laser line at 488 nm and helium–neon-laser lines at 543 and 633 nm and appropriate filter configurations, such as the Leica SP2 or Zeiss Pascal Axioplan LSM.
31. To be able to visualize vesicular structures at least a 63× immersion oil objective is recommended.

3.3. Co-localization of Internalized DNA and Transferrin or Dextran

1. Follow the procedure in **Subheading 3.2.** However, in **step 4** just prior to the addition of the ligand solution, add 100 μg/mL transferrin-AlexaFluor-647 or 300 μg/mL dextran-Alexa Fluor 647 to the peptide/DNA mixture to visualize clathrin-dependent endocytosis or macropinocytosis, respectively. Analyze the cells either live or fixed and counter-stained. If the cells are fixed and counter-stained, omit adding TO-PRO-3 under **step 18**, as the fluorescence of TO-PRO-3 is in the far-red spectrum just as Alexa Fluor 647.

3.4. Co-localization of Internalized DNA and Caveolin-1-YFP

1. Seed 3×10^6 HeLa cells in a 75-cm^2 flask.
2. Allow the cells to grow for approx. 24 h.

3. Rinse the cell layer with 15 mL PBS.
4. Detach the cells with 3 mL Trypsin/EDTA for approx. 2 min or until the cells begin to detach.
5. Add 10 mL of DMEM with 10% FCS and suspend the cells.
6. Count an aliquot of the cells using a Bürker chamber.
7. Pellet the cells by centrifugation at $300 \times g$ for 5 min at 4°C.
8. Wash the cells twice in ice-cold PBS with centrifugation between the washes.
9. Suspend the cell pellet in 1 mL ice-cold PBS (in total $6-8 \times 10^6$ cells).
10. Transfer 500 μL of the cell suspension to a 4-mm electroporation cuvette (the remaining cell suspension can be used as non-transfected control cells or for other transfections; *see* **Note 9**).
11. Add up to 50 μg of caveolin-1-YFP encoding plasmid to the cell suspension in the electroporation cuvette (*see* **Note 10**).
12. Add PBS to bring the final volume to 700 μL.
13. Keep the cells on ice for 30 min.
14. Resuspend the cells by tapping on the cuvette.
15. Wipe the cuvette dry using a paper towel, avoid putting any fingers on the metal surfaces of the cuvette.
16. Electroporate the cells at 320 V using the low voltage setting of the electroporator (*see* **Note 11**).
17. Immediately, add 1 mL of DMEM with 10% FCS.
18. Resuspend the cells gently using a pipette.
19. Dilute an appropriate volume of the cell suspension containing 350,000 cells in 5 mL DMEM with 10% FCS.
20. Seed 500 μL/well of the diluted cell suspension in an 8-well chamber slide.
21. Allow the cells to grow for 18–48 h. Transfection efficiency can be monitored using an inverted fluorescence microscope, if desired.
22. Prepare the cells for analysis either as live cells or as fixed and counter-stained as described under **Subheading 3.2.** However, as the YFP tagged caveolin-1 will fluoresce in the green spectrum the use of Alexa Fluor 546 or Alexa Fluor 647 (or similar) labeled DNA is necessary.

3.5. Disruption of Dynamin-Dependent Endocytosis Using Dominant-Negative Dynamin

This protocol requires fixation and permeabilization of the cells, therefore it is only applicable to ligands for which this is possible. GFP-tagged dominant-negative dynamin constructs are, however, available *(19)* and should enable the use of live cells.

1. Transfect cells according to **Subheading 3.4**. In **step 10**, however, transfer the cell suspension to two electroporation cuvettes, 500 μL each. Then in **step 11** add 50 μg HA-dyn2-WT and HA-dyn2-K44A to the two cuvettes, respectively.

Proceed with the electroporation as in **Subheading 3.4, steps 12–21** for each of the constructs (*see* **Note 12**).

2. Following 18–48 h of incubation wash the cells once with DMEM.
3. Add 150 µL of pre-warmed peptide/DNA solution.
4. Incubate the cells for 2 h at 37°C.
5. When 30 min of the incubation remains: Add 150 µL of pre-warmed DMEM containing 100 µg/mL transferrin-Alexa Fluor 647 to one well with cells of each treatment.
6. Remove the slide from the incubation chamber and place on ice, allow to cool for 15 min.
7. Add ice-cold peptide/DNA solution to one well for the 4°C incubation control.
8. Incubate for 30 min on ice in the dark.
9. Remove all incubation solutions.
10. Wash with 250 µL ice-cold PBS.
11. Wash once with 250 µL/well of 1 *M* NaCl in PBS, do not allow the cells to remain in the hyperosmotic buffer for more than 30 s.
12. Quickly remove the 1 *M* NaCl buffer and gently wash twice with PBS.
13. Remove the PBS and add 200 µL of freshly prepared or thawed 4% paraformaldehyde, incubate for 15 min.
14. Wash twice with PBS.
15. Add 200 µL PBS with 0.1% Triton X-100, incubate for 10 min.
16. Wash three times with PBS.
17. Block the samples overnight with 300 µL PBS with 2% BSA.
18. Remove blocking solution and add 100 µL of anti-HA antibody (1:100) diluted in PBS+2% BSA and incubate for 1 h at RT.
19. Wash three times with PBS with 2% BSA.
20. Add 100 µL of Alexa Fluor 488-conjugated anti-mouse antibody (1:500) diluted in PBS+2% BSA and incubate for 1 h at RT.
21. Wash three times with PBS with 2% BSA.
22. Add 100 µL of PBS+2% BSA with phalloidin-Alexa Fluor 543 (1:100).
23. Wash three times with PBS, remove the PBS.
24. Remove the plastic wells from the glass slide using the supplied tools (Nunc).
25. Air-dry the slide by waving it in the air.
26. When the slide is almost dry, add two small drops of permafluor on different parts of the slide.
27. Carefully place a 60 × 24 mm cover glass on the slide and let the permafluor solution spread out by capillary force.
28. Keep the slide in the fridge overnight to allow the permafluor to solidify. To avoid the slide from drying, seal the slit between the cover glass and the slide by applying a thin layer of nail polish over the slit. Avoid any nail polish on the surface above or below the samples.
29. Analyze the slide using a confocal microscope system equipped with an argon-laser line at 488 nm and helium–neon-laser lines at 543 and 633 nm and appropriate filter configurations, such as the Leica SP2 or Zeiss Pascal Axioplan LSM.

4. Notes

1. In this chapter, we provide protocols that have been optimized for the study of HIV-tat-mediated DNA uptake. These protocols are, however, also intended to be applicable to studies with other peptides and cargos.

2. Plastic and special cell culture-treated glass slides can give rise to high background staining from the fluorescently labeled ligands, due to unspecific binding of the ligand to the glass or coating. If you wish to minimize the unspecific background stain, we recommend using an untreated glass slide, such as LabTek (Nunc).

3. Endocytosis interfering drugs can be included in the medium 30 min to 1 h prior to ligand incubation and potentially also during ligand incubation. Examples of drugs that we and others have used include the following: nystatin and methyl-cyclodextrin for membrane cholesterol depletion; cytochalasin D, latrunculin A, and jasplakinolide for actin filament disruption; chlorpromazine and hypertonic shock for clathrin-mediated endocytosis interference; amiloride and di-methyl-amiloride for interference with macropinocytosis through Na^+/H^+ antiport inhibition.

4. For FACS experiments, DNA concentrations between 1 and 10 µg/mL are suggested with peptide to DNA ratios between 0 and 10 (w/w). For microscopy experiments, we recommend using higher DNA concentrations (5–20 µg/mL) and similar peptide to DNA ratios as those used in the flow cytometry experiments.

5. We suggest time points between 10 min and 8 h for initial screening experiments.

6. Endocytotic processes are energy- and temperature-dependent, therefore experiments carried out at 4°C during ligand incubation are proper controls to determine whether the ligand uptake is likely to be endocytic and/or there are any remaining surface associated ligands after the washing procedures. Initial flow cytometry experiments should furthermore evaluate the time-dependence and concentration-dependence of the ligand uptake, as this will aid the design of more qualitative studies as well as characterizing fundamental properties of the uptake.

7. One of the major drawbacks with the use of flow cytometry is its inability to discriminate between internalized and cell surface associated ligands. In order to minimize the risk of misinterpreting the origin of the fluorescent signal, stringent cell surface washing and stripping procedures are required. Trypsin treatment of the intact cells prior to flow cytometry will remove most surface-bound ligands. Concentrated salt washes (e.g., 1 M NaCl in PBS) are recommendable, if the receptor/ligand interaction is primarily electrostatic in nature (as is the case for most CPPs), and trypsin treatment must be avoided.

8. Microscopy slides are ideally analyzed within a few days of preparation. The slides can, however, be stored in the dark at 4°C for weeks or months with acceptable staining quality.

9. In any experiment, where cells are to be transfected prior to analysis of DNA uptake (peptide, protein-mediated or similar) we do not recommend the use of

cationic lipids (e.g., Lipofectamine) as transfection agents. These compounds may be difficult to be completely removed from cells prior to the addition of CPP–DNA complexes and will thus influence the DNA uptake activity. We have found that, electroporation of cells 18–48 h prior to incubation with DNA complexes does not influence CPP-mediated DNA delivery. It is, however prudent, at least during initial experiments, to compare the behavior of the transfected cells to that of non-transfected control cells.

10. Plasmid DNA (10–50 µg) should be used for each transient electroporation. In general, a higher DNA concentration will result in higher transfection efficiency.

11. We use the low voltage setting of the BTX ECM399 (Genetronics) electroporator, which provides a 1050-mF electroporation pulse. The electronic properties of the electroporator will depend on the specific model and manufacturer, therefore the settings given here may not be optimal when using other equipment. As a rule of thumb, a pulse intensity resulting in 30–50% cell death the day after the transfection will result in optimal transfection efficiency. Optimal transfection conditions are ideally found using a plasmid encoding a fluorescent protein (e.g., GFP) or a fluorescent fusion protein such as the caveolin-1-YFP construct used in this chapter.

12. At least 3 wells/construct (no ligand, transferrin and peptide/DNA mixture) is necessary. Non-transfected cells, both with and without ligand should also be included in the experiments.

References

1. Vives, E., Brodin, P., and Lebleu, B. (1997) A truncated HIV-1 Tat protein basic domain rapidly translocates through the plasma membrane and accumulates in the cell nucleus. *J Biol Chem* **272,** 16010–16017.
2. Derossi, D., Calvet, S., Trembleau, A., Brunissen, A., Chassaing, G., and Prochiantz, A. (1996) Cell internalization of the third helix of the Antennapedia homeodomain is receptor-independent. *J Biol Chem* **271,** 18188–18193.
3. Lindgren, M., Hallbrink, M., Prochiantz, A., and Langel, U. (2000) Cell-penetrating peptides. *Trends Pharmacol Sci* **21,** 99–103.
4. Richard, J. P., Melikov, K., Vives, E., Ramos, C., Verbeure, B., Gait, M. J., Chernomordik, L. V., and Lebleu, B. (2003) Cell-penetrating peptides: a re-evaluation of the mechanism of cellular uptake. *J Biol Chem* **278,** 585–590.
5. Richard, J. P., Melikov, K., Brooks, H., Prevot, P., Lebleu, B., and Chernomordik, L. V. (2005) Cellular uptake of unconjugated TAT peptide involves clathrin-dependent endocytosis and heparan sulfate receptors. *J Biol Chem* **280,** 15300–15306.
6. Ferrari, A., Pellegrini, V., Arcangeli, C., Fittipaldi, A., Giacca, M., and Beltram, F. (2003) Caveolae-mediated internalization of extracellular HIV-1 tat fusion proteins visualized in real time. *Mol Ther* **8,** 284–294.
7. Wadia, J. S., Stan, R. V., and Dowdy, S. F. (2004) Transducible TAT-HA fusogenic peptide enhances escape of TAT-fusion proteins after lipid raft macropinocytosis. *Nat Med* **10,** 310–315.

8. Rejman, J., Oberle, V., Zuhorn, I. S., and Hoekstra, D. (2004) Size-dependent internalization of particles via the pathways of clathrin- and caveolae-mediated endocytosis. *Biochem J* **377,** 159–169.

9. Sandgren, S., Cheng, F., and Belting, M. (2002) Nuclear targeting of macromolecular polyanions by an HIV-Tat derived peptide. Role for cell-surface proteoglycans. *J Biol Chem* **277,** 38877–38883.

10. Suzuki, T., Futaki, S., Niwa, M., Tanaka, S., Ueda, K., and Sugiura, Y. (2002) Possible existence of common internalization mechanisms among arginine-rich peptides. *J Biol Chem* **277,** 2437–2443.

11. Belting, M. (2003) Heparan sulfate proteoglycan as a plasma membrane carrier. *Trends Biochem Sci* **28,** 145–151.

12. Mislick, K. A., and Baldeschwieler, J. D. (1996) Evidence for the role of proteoglycans in cation-mediated gene transfer. *Proc Natl Acad Sci U S A* **93,** 12349–12354.

13. Shieh, M. T., WuDunn, D., Montgomery, R. I., Esko, J. D., and Spear, P. G. (1992) Cell surface receptors for herpes simplex virus are heparan sulfate proteoglycans. *J Cell Biol* **116,** 1273–1281.

14. Patel, M., Yanagishita, M., Roderiquez, G., Bou-Habib, D. C., Oravecz, T., Hascall, V. C., and Norcross, M. A. (1993) Cell-surface heparan sulfate proteoglycan mediates HIV-1 infection of T-cell lines. *AIDS Res Hum Retroviruses* **9,** 167–174.

15. Hopkins, C. R., Miller, K., and Beardmore, J. M. (1985) Receptor-mediated endocytosis of transferrin and epidermal growth factor receptors: a comparison of constitutive and ligand-induced uptake. *J Cell Sci Suppl* **3,** 173–186.

16. Glebov, O. O., Bright, N. A., and Nichols, B. J. (2006) Flotillin-1 defines a clathrin-independent endocytic pathway in mammalian cells. *Nat Cell Biol* **8,** 46–54.

17. Pelkmans, L., Kartenbeck, J., and Helenius, A. (2001) Caveolar endocytosis of simian virus 40 reveals a new two-step vesicular-transport pathway to the ER. *Nat Cell Biol* **3,** 473–483.

18. Damke, H., Binns, D. D., Ueda, H., Schmid, S. L., and Baba, T. (2001) Dynamin GTPase domain mutants block endocytic vesicle formation at morphologically distinct stages. *Mol Biol Cell* **12,** 2578–2589.

19. Lee, E., and De Camilli, P. (2002) Dynamin at actin tails. *Proc Natl Acad Sci U S A* **99,** 161–166.

8

Selection and Characterization of Antibodies from Phage Display Libraries Against Internalizing Membrane Antigens

Johan Fransson and Carl A.K. Borrebaeck

Summary

Macromolecular delivery systems require high target cell specificity and efficient intracellular uptake. Monoclonal antibodies (mAbs) have been shown to successfully meet these needs and should, due to their biological nature and thus minimal toxicity and limited immunogenicity, be optimal delivery vehicles for various macromolecules (e.g., toxins, drugs, oligonucleotides). Such antibodies could be retrieved from phage display libraries by carefully designed selection and screening methods. In this chapter, we provide protocols for the isolation of phage-derived antibodies reactive to cell surface receptors, which upon binding will induce receptor-mediated internalization of the antibody/receptor complexes. In addition, a protocol describing the identification of target antigens by immunoprecipitation (ip) of cell lysates and preparation of gel plugs for subsequent MALDI-TOF analysis is included. Furthermore, we suggest several techniques that could be employed to confirm the specificity as well as the drug delivery potential of isolated clones.

Key Words: Phage display; Internalization; Receptor-mediated endocytosis; Antibody; Antigen identification; Immunoprecipitation; Receptor; Membrane protein.

1. Introduction

The inherent features of monoclonal antibodies (mAbs), i.e., high specificity and generally low toxicity, make them ideal for targeting transformed cell surfaces. Upon binding, the mAbs could induce homo- or hetero-dimerization of the receptors, which in turn stimulate internalization into the target cells

From: *Methods in Molecular Biology, vol. 480: Macromolecular Drug Delivery*, Edited by: M. Belting
DOI 10.1007/978-1-59745-429-2_8, © Humana Press, a part of Springer Science+Business Media, LLC 2009

by various mechanisms, usually divided into clathrin-dependent or clathrin-independent endocytosis *(1,2)*. Regardless of the route of endocytosis, the internalization of antibodies makes them highly suitable delivery vehicles for macromolecules and drugs *(3)*.

Phage display, originally introduced to the scientific community by Smith in 1985 *(4)*, is a well-established technique for the selection of protein, e.g., antibodies, or peptide binders to various substrates *(5,6)*. A selection is basically carried out by mixing the phage library with the antigen, where the incubation time and physical parameters, e.g., temperature and pH, are tuned to fit the purpose of the selection. After the incubation, unbound phages are discarded by a series of washing steps, followed by the retrieval of bound phages during a final elution, e.g., by low pH buffer, trypsin treatment or competitively by adding excess antigen *(7)*. The preferable antigen format when selecting antibodies from phage display libraries is a purified polypeptide. In this case, the antigen can be immobilized onto a solid support or captured on streptavidin-coated beads *(8)*. Disadvantages with this system include the risk of denaturing the antigen during the immobilization step and that the antibody only recognizes its antigen when coupled to the carrier. The latter is commonly seen when selecting antibodies against small molecules, such as haptens *(9)*.

However, in the case of therapeutic antibody/antigen discovery, selection schemes should be chosen to improve the probability of finding binders that mediate functional effects on target cells. As a first step toward the generation of functional antibodies, whole cells should be used as the antigen source *(10–15)*. This will ensure the native configuration of the antigen, which is present in its original cellular context. Interestingly, to successfully isolate antibodies to some antigens, cellular association is a prerequisite *(16)*. To reduce the number of non-specific binders, the phage library is usually pre-incubated with a subtractor cell line, not expressing target antigen(s). Furthermore, if selections are carried out under more physiological conditions, i.e., at 37°C, bound phages are allowed to internalize into the cells of interest *(17–22)*. Internalized phages can be rescued after cell lysis. Depending on the aim of the study, selections are either carried out on a cell line stably expressing the antigen of interest, or done without prior knowledge of the nature of the isolated specificities, thus allowing for target discovery. When novel antigens are sought after, the panning and screening steps need to be followed by an antigen identification step, usually carried out by immunoprecipitation (ip) of cell lysates and subsequent mass spectrometry analysis. In the following sections, we will provide detailed protocols for the retrieval of phage antibodies that induce internalization upon cell surface binding, a necessary function for an intracellular delivery system. Furthermore, screening of selected clones for surface binding, using flow cytometry, and for internalization, using confocal microscopy, will also be described. In addition, we suggest protocols for antigen identification by immunoprecipitation and MALDI-TOF analysis. Finally, an indirect immunotoxin cytotoxicity protocol is provided to assess the macromolecular delivery potential of selected clones without time consuming toxin conjugation steps. Importantly, the protocols described in **Subheadings 3.5.–3.7.** require reformatting of scFv phage clones to full IgG format, the procedure for which is not provided herein, but can be found elsewhere, please *see* refs. *(23,24)*.

2. Materials
2.1. Phage Library Preparation

1. *Escherichia coli* strain TOP10 F′ (Invitrogen)
2. Bacterial culture flasks
3. Corning 50 mL centrifuge tubes (Sigma, cat # CLS430828)
4. Medium, 2 × YT/amp/glu: 2 × YT medium containing 100 μg/mL ampicillin and 2% glucose
5. YT/amp/glu agar plates: 100 mm YT agar plates containing 100 μg/mL ampicillin and 2% glucose
6. Medium, 2 × YT/amp/kan/IPTG: 2 × YT medium containing 100 μg/mL ampicillin, 25 μg/mL kanamycin, and 100 μM isopropyl β-D-thiogalactopyranoside (IPTG)
7. Helper phage: VCSM13, R408, M13K07 (Stratagene, New England Biolabs, or GE Healthcare)
8. Phage library glycerol stock (*see* **Note 1**)
9. Precipitation solution: 25% PEG6000/2.5 *M* NaCl. First, dissolve 125 g PEG6000 in approx. 200 mL sterile distilled water. Make 2.5 *M* NaCl by dissolving 73 g NaCl in 500 mL sterile distilled water. Add 2.5 *M* NaCl to the PEG6000 solution to a final volume of 500 mL. To obtain a homogenous solution, heating of the mixture might be required. Filter the final solution through a 0.2 μm filter.

2.2. Whole Cell Phage Selection
2.2.1. Negative Subtractor Cell Pre-selection

1. Corning 50 mL centrifuge tubes (Sigma, cat # CLS430828)
2. General purpose centrifuge
3. Subtractor cell line of choice (*see* **Note 2**), with recommended culture medium
4. Wash medium: RPMI 1640 cell culture medium (Sigma), 10% (v/v) foetal calf serum (Invitrogen), 50 m*M* HEPES buffer, pH 7.0 (Sigma), 2 m*M* EDTA
5. Phage antibody library, prepared as described in **Subheading 3.1.**
6. Precipitation solution, prepared as described in **Subheading 2.1.**

2.2.2. Positive Target Cell Selection

1. Heraeus CO_2 Cell Culture Incubator
2. Corning 50 mL centrifuge tubes (Sigma, cat # CLS430828)
3. Target cell line of choice (*see* **Note 2**), with recommended culture medium
4. Wash medium, prepared as described in **Subheading 2.2.1.**
5. Phage antibody library, prepared as described in **Subheading 3.1.**
6. Precipitation solution, prepared as described in **Subheading 2.1.**
7. Ficoll (40% v/v, GE Healthcare), 2% (w/v) bovine serum albumin (BSA, Sigma) in phosphate buffered saline (PBS) without Mg^{2+}/Ca^{2+} (Invitrogen)
8. Stripping buffer: 0.1 M glycine, pH 2.2. Dissolve glycine in distilled ultra pure water and adjust pH to 2.2 using 6 M HCl.

2.3. Screening for Cell Surface Binding by Flow Cytometry

1. Flow cytometry instrumentation (e.g., FACScan, FACSCalibur, FACSArray, from BD Biosciences)
2. Trypsin/EDTA (cell culture tested, Sigma)
3. PBS (cell culture tested, Sigma)
4. Target cell line
5. Subtractor cell line
6. FACS buffer: 0.5% (w/v) BSA in PBS
7. Mouse anti-M13 antibody (GE Healthcare, cat # 27-9420-01)
8. Suitable secondary antibody reagent, e.g., phycoerythrin-conjugated Goat $F(ab')_2$ anti-Mouse antibody (R&D Systems).

2.4. Verification of Internalization by Confocal Microscopy

1. CultureWell chambered coverglass for cell culture (Invitrogen, cat # C37000)
2. Stripping buffer: 0.1 M glycine/HCl pH 2.2, prepared as described in **Subheading 2.2.2.**
3. BSA 1% (w/v) in PBS
4. Paraformaldehyde solution, 16% (formaldehyde solution) EM Grade (Electron Microscopy Sciences, cat # 15710)
5. Image-iT FX signal enhancer (Invitrogen, cat # I36933)
6. Goat serum (Sigma, G 9023), passed through a syringe filter to remove particulate material
7. Triton X-100 0.1% (Sigma, cat # 234729) or 0.2% saponin (Sigma, cat # 4521) diluted in PBS
8. ProLong Gold antifade reagent with 4',6-diamidino-2-phenylindole DAPI (Invitrogen, cat # P36931)
9. Mouse anti-M13 antibody (GE Healthcare, cat # 27-9420-01)
10. Alexa Fluor 488 goat anti-mouse IgG (H + L) 2 mg/mL (Invitrogen, cat # A11029)
11. BioRad confocal microscope (Hercules, CA)

2.5. Antigen Identification by Immunoprecipitation

1. Selected antibody clones reformatted into fully human IgG1 format [prepared as described elsewhere, *see* refs. *(23,25)*]
2. Lysis buffer: 0.5% NP40 (Sigma), 25 m*M* Tris–HCl, pH 7.5, 5 m*M* EDTA, 150 m*M* NaCl, protease inhibitor cocktail (Roche) (*see* **Note 3**)
3. Wash buffer: 0.1% NP40, 25 m*M* Tris–HCl, pH 7.5, 5 m*M* EDTA, 150 m*M* NaCl, protease inhibitor cocktail
4. Hyptonic buffer: 5 m*M* Tris–HCl, pH 7.5, 5 m*M* EDTA, protease inhibitor cocktail
5. Dounce homogenizer
6. Immunoprecipitation Starter Kit, containing Protein A and Protein G Sepharose beads (GE Healthcare)
7. Conventional table-top centrifuge
8. Ultracentrifuge

2.6. Preparation of Samples for Mass Spectrometry

1. Acetonitrile (ACN) HPLC grade (Pierce, cat # 51101)
2. Ammonium bicarbonate (Sigma, cat # 09830)
3. SpeedVac System (Savant)
4. No-weight dithiothreitol (DTT) (Pierce, cat # 20291)
5. Iodoacetamide (Sigma, cat # 57670)
6. Trypsin Gold, Mass Spectrometry Grade (Promega, cat # V5280)
7. Trifluoroacetic acid (TFA) (Pierce, cat # 28901)
8. MALDI target plate (Applied Biosystems, Foster City, CA, Cat no. V700668#)
9. α-Cyano-4-hydroxy cinnamic acid (CHCA) (Applied Biosystems)
10. Applied Biosystems Maldi Workstation 4700

2.7. Assessment of Macromolecular Drug Delivery Potential

1. Selected antibody clones reformatted into fully human IgG1 format [prepared as described elsewhere, *see* refs. *(23,25)*].
2. Saporin-conjugated anti-human antibody (Hum-ZAP, Advanced Targeting Systems, San Diego, CA)
3. [^{3}H]thymidine (GE Healthcare)
4. MicroBeta Liquid Scintillation Counter

3. Methods
3.1. Phage Library Preparation

1. Inoculate 2-L culture flasks containing 200 mL of YT/amp/glu medium with a phage library glycerol stock.
2. Incubate at 37°C with shaking (250 rpm), until the culture has reached an OD (600 nm) of 0.5.
3. Infect the culture with helper phage. Incubate for 30 min at 37°C without shaking.

4. Centrifuge the bacterial cultures at 3000 × g for 10 min. Discard the supernatant.
5. Resuspend the pellet in 2 × YT/amp/kan/IPTG medium and incubate in 2-L cell culture flasks with shaking overnight at 30°C.
6. Centrifuge the bacteria at 3000 × g for 10 min. Transfer the supernatant containing the produced phage to new bottles and precipitate phages by adding 25% PEG6000/2.5 *M* NaCl to the phage solution in a ratio of 1:4. Incubate for 4 h or overnight at 4°C.
7. Pellet the phages by centrifugation at 4°C for 30 min at 15,000 × g.
8. Resuspend the pellet in suitable volume of PBS (suggested volume 1–2 mL).
9. Determine phage titer by making 1/10 dilution series of the phage solution in PBS. Estimate the titer to be between 10^{10} and 10^{13} cfu/mL. Add 10 µL of the diluted phage from, e.g., the 10^8, 10^{10}, 10^{12} dilution to 190 µL of exponentially growing *E. coli* TOP10F′. Incubate without shaking for 30 min at 37°C. Plate 100 µL of the infected bacterial culture onto YT/amp/glu plates and incubate at 37°C overnight. Calculate the titer by counting the number of colonies and multiply by the dilution factor. E.g., 200 colonies on the plate from the 10^8 dilution gives a titer of $200 \times 10^8 \times 20$ (10 µL of phage was added to 190 µL of bacteria) = 4000×10^8 cfu/100 µL = 4×10^{12} cfu/mL.

3.2. Whole Cell Phage Selection

3.2.1. Negative Subtractor Cell Pre-selection

1. Pellet subtractor non-target cells (10–500×10^6 cells) by centrifugation at 4°C ($400 \times$ g, 5 min).
2. Dissolve the pelleted cells in wash medium and add the phage library (1×10^{13} cfu total phage). Adjust to a final volume of 4 mL. Incubate the cell/phage mixture at 4°C for 3 h on rotation.
3. Pellet the cells as described above.
4. Collect the supernatant containing the unbound phages.
5. Dissolve the pelleted cells in 4 mL wash medium
6. Pellet the cells again and pool the supernatant with the supernatant from **step 4**.
7. Precipitate the phages as described in **step 6** in **Subheading 3.1.**
8. Pellet the phages by centrifugation at 4°C, 30 min, at 20,000 × g.
9. Discard the supernatant and dissolve the pellet in 1 mL of wash medium and store at 4°C until further use.

3.2.2. Positive Target Cell Selection

1. Pellet target cells (10×10^6 cells) by centrifugation at 4°C ($400 \times$ g, 5 min).
2. Dissolve the pelleted cells by adding the 1 mL solution containing the pre-selected library from **step 9** in **Subheading 3.2.1.** Add another 1 mL wash medium to the tube that contained the pre-selected library to wash out the remaining phages. Incubate the cell/phage mixture at 4°C for 1 h on rotation.

3. To allow internalization of bound phages, transfer the phage/cell suspension to a humidified atmosphere, containing 5% CO_2, and incubate at 37°C for 1 h (*see* **Note 4**).

4. Pellet the cells by centrifugation (as described in **step 1**) and resuspend the pellet in 1 mL wash buffer.

5. Transfer the cell suspension to a 50-mL centrifuge tube containing 10 mL of 40% Ficoll, 2% BSA/PBS (without Ca^{2+}) and centrifuge as described above.

6. Resuspend the pellet in 1 mL PBS (with Ca^{2+}) and add PBS to a final volume of 10 mL.

7. Pellet the cells as described above.

8. Strip surface-bound phages by adding 5 mL stripping buffer and incubate for 15 min.

9. Pellet the cells as described above.

10. Lyse the cell pellet by resuspending in 1 mL of 100 mM triethylamine and incubate for 5 min at room temperature.

11. Neutralize the lysate with 100 μL 1 M Tris–HCl, pH 8.3.

12. To rescue the selected phage, infect a TOP10F′ culture (OD_{600} = 0.5) by the output phages from the selection (30 min, 37°C), spin down the cells, and resuspend in 1 mL of the supernatant. Plate the cell suspension on agar plates (amp/tet) and incubate at 37°C overnight.

13. Harvest the cells from the plates and suspend in 2× YT/amp/glu media and store with 15% glycerol at –80°C.

14. Prepare new phage stocks from such pools of bacteria and repeat the selection two to four times, depending on the output/input ratios.

15. After the last selection, pick individual colonies, grow in culture, and store as monoclonal glycerol stocks at –80°C.

3.3. Screening for Cell Surface Binding by Flow Cytometry

1. Cells that should be used for screening (subtractor and target cell lines) are grown to confluency in their appropriate growth medium.

2. Wash cells in PBS.

3. If adherent, detach cells from culture flasks by incubation with 1× trypsin/EDTA or, if the target antigen is sensitive to tryptic digestion, only EDTA.

4. Quench the protease activity by adding growth medium.

5. Count the cells and wash with PBS by centrifugation at 400 × g, 5 min, at 4°C.

6. Add 2×10^5–1×10^6 cells/staining reaction to 5 mL FACS tubes or to V-bottom 96-well plates for FACSArray.

7. Add monoclonal phage (concentration of amplified phage should be between 1×10^{10} and 1×10^{13} cfu/mL) at 1/10 dilution (*see* **Note 5**).

8. Incubate for 1 h at 4°C.

9. Wash cells as described in **step 5**.

10. Add mouse anti-M13 (pVIII) antibody at a 10 μg/mL final concentration and incubate for 30 min at 4°C.

11. Wash cells as described in **step 5**.

12. Add secondary antibody at a final concentration of 10 μg/mL and incubate for 30 min at 4°C.
13. Wash cells as described in **step 5**, resuspend cell pellet in 500 μL (for analysis in 5 mL FACS tubes) or 100 μL (for 96-well plate format).
14. Analyze binding of selected phage clones by flow cytometry.

3.4. Verification of Internalization by Confocal Microscopy

1. Grow target cells (20,000–50,000 cells/well) overnight in culture slides with detachable chambers.
2. Incubate cells in fresh growth medium for 1 h.
3. Add phages to the chambers and incubate at 37°C for 1 h. As negative control for internalization, incubate a control slide with phages at 4°C for 1 h followed by two washes in ice cold PBS. Then, proceed to **step 7** below.
4. Wash wells twice with 37°C PBS.
5. Strip cell surface bound phages by washing twice in stripping buffer for 10 min.
6. Wash wells twice with 37°C PBS.
7. Fix cells in 4% (w/v) formaldehyde (diluted in PBS buffer) for 40 min at 4°C.
8. Rinse each sample in PBS buffer —three to four times.
9. Permeabilize cells in 0.1% Triton X-100 or 0.2% saponin for 15 min.
10. Rinse each sample in PBS buffer —three to four times.
11. Apply four drops (approx. 200 μL) of Image-iT FX signal enhancer. Incubate for 30 min at room temperature in a humid environment.
12. Rinse thoroughly with PBS buffer.
13. Incubate cells with mouse anti-M13 antibody (diluted in 2% goat serum/1%BSA/PBS for 1 h at room temperature.
14. Wash three times with PBS.
15. Add secondary Alexa Fluor 488 goat anti-mouse antibody (diluted in PBS, recommended dilution 1:200–1:1000) and incubate for 1 h at room temperature.
16. Wash sample three times with PBS.
17. Remove ProLong Gold antifade reagent from freezer and allow warming to room temperature. Remove any liquid from specimen and apply one drop of ProLong Gold antifade reagent. Cover slide-mounted specimen with a coverslip (provided with the CultureWell chambers).
18. Analyze slides with a confocal microscope.

3.5. Antigen Identification by Immunoprecipitation

1. Wash 50–500 × 10⁶ target cells twice in 50 mL PBS.
2. Suspend the pellet in 5 mL hypotonic buffer. Incubate on ice for 15 min to swell the cells.
3. Homogenize in a glass Dounce homogenizer on ice.
4. Pellet nuclei and unbroken cells at 1000 g for 5 min. Save supernatant.
5. Resuspend pellet in 5 mL hypotonic buffer and homogenize a second time.
6. Centrifuge at 1000 × g for 5 min and combine the supernatant with that from **step 4**. Save pellet for solubilization in lysis buffer overnight.

7. Centrifuge the 1000 × g supernatants at 10,000 × g for 12 min. Save pellet for solubilization in lysis buffer overnight.
8. Centrifuge the 10,000 × g supernatant for 40 min at 100,000 × g.
9. Solubilize the membrane proteins in the 100,000 × g pellet in lysis buffer for 15 min on ice with intermittent vortexing.
10. Clear the solubilized proteins by another 100,000 × g centrifugation for 30 min.
11. Add Protein A Sepharose to pre-clear the cell lysates (100 μL Protein A/mL lysate) for 30 min.
12. Add antibody of unknown specificity to the pre-cleared cell lysates. Incubate for 2 h at 4°C on rotation (*see* **Note 7**).
13. Add 50 μL of Protein A Sepharose to the antibody/cell lysate suspension. Incubate for 30 min at 4°C on rotation.
14. Add antibody–Protein A Sepharose complex to pre-cleared cell lysate. Incubate for 2 h on ice, on rotation.
15. Wash the immune complexes in wash buffer ten times.
16. Resuspend the last pellet in SDS sample buffer and run on 4–12% SDS-PAGE.

3.6. Preparation of Samples for Mass Spectrometry

3.6.1. Destaining/Dehydration/Rehydration

1. Wash gel plugs in 50% ACN in 25 mM NH$_4$HCO$_3$ (3 × 20 min).
2. Dry gel plugs in SpeedVac for 15 min.

3.6.2. Reduction/Alkylation

1. Make fresh 10 mM DTT in 100 mM NH$_4$HCO$_3$.
2. Make fresh 55 mM iodoacetamide in 100 mM NH$_4$HCO$_3$.
3. Add 25 μL of 10 mM DTT to the dried SpeedVac sample.
4. Incubate at 56°C for 1 h.
5. Discard the DTT and add 25 μL of 50 mM iodoacetamide, incubate for 45 min at room temperature. Keep in the dark, since the iodoacetamide is light sensitive.

3.6.3. Washing

1. Discard the iodoacetamide, add 50 μL NH$_4$HCO$_3$, and put on vortex for 10 min at room temperature.
2. Add new 50 μL NH$_4$HCO$_3$ and vortex again for 10 min.
3. Wash samples in 50% ACN followed by vortex for 10 min.
4. Dry samples in SpeedVac for 15 min.

3.6.4. Tryptic Digestion

1. Add 15 μL trypsin (15 ng/μL in 25 mM NH$_4$HCO$_3$).
2. Incubate overnight at 37°C in a heated water bath. The gel pieces will swell and later the solution will condensate on the tube lid.

3. Aspirate condensed liquid from the lid and add 15 μL 50% ACN, 1% TFA, and incubate on a shaker for 10 min.
4. Extracted peptides are then aspirated and saved in new tubes at −20°C until later. The gel pieces should also be saved if additional extraction is needed.

3.6.5. Spotting on Target Plates

1. Clean a MALDI target plate carefully. Wash sequentially in ethanol, distilled ultra pure water, soap, ethanol, and distiller water. Dry and polish the plate with polish.
2. Spot 1 μL/sample in the plate. Let the spot dry.
3. Make the matrix by weighing 5 mg of CHCA and dissolve in 100 μL of 75% ACN, 1% TFA. This step should be carried out in tubes made of glass. If plastic tubes are used, softeners from the plastic might be dissolved due to the high concentration of ACN.
4. Add 0.5–1 μL matrix and ensure that crystals are formed that cover the entire sample well. Submit samples for analysis by mass spectrometry and compare peptide fingerprint with theoretic peptide masses in online databases (*see* **Note 8**).

3.7. Assessment of Macromolecular Drug Delivery Potential

1. Seed target cells, and non-target cells for negative control, at 10,000/well (100,000/mL) in 96-well plates and incubate overnight at 37°C in 5% CO_2.
2. Add 200 ng/well of primary antibody.
3. Treat half of the test wells with 100 ng/well of saporin-conjugated anti-human antibody (Hum-ZAP).
4. Assess the cytotoxicity of the immunotoxin conjugate by [^{3}H]thymidine incorporation assay. Pulse cells after 56 h by the addition of 0.5 μCi/well of [^{3}H]thymidine and incubate for another 16 h.
5. Determine the incorporation of [^{3}H]thymidine by reading in Beta Liquid Scintillation Counter.

4. Notes

1. The antibody phage library could be constructed or purchased from commercial vendors. The method for constructing an antibody phage library is beyond the scope of this chapter, but can be found elsewhere, please *see* refs. *(24,26–29)*. Commercially available libraries could be purchased and licensed from, e.g., Dyax (www.dyax.com), New England Biolabs (www.neb.com), and BioInvent International AB (www.bioinvent.com). When making a decision of the source and format of the phage display system, keep in mind the two general types of libraries to chose from; those cloned into phagemid vectors and those cloned into phage vectors. The display of antibody fragments using phage vectors results in the antibody gene becoming a part of the phage genome and essentially all phages will display the antibody in 3–5 copies/phage *(30)*. In the phagemid system, the pIII/antibody

fusion protein is encoded on a phagemid that lack genes necessary to make functional phage. Instead, these genes are provided by a helper phage *(31)*, which when introduced into phagemid-carrying bacteria will produce phages mostly displaying antibody fragments monovalently. The use of either phage or phagemid vectors will have consequences on the nature of the retrieved antibodies. Owing to the avidity effect created by multivalent display, phage display (in comparison to phagemid display) will generally result in the isolation of clones with lower affinity *(32)* and subsequent affinity maturation steps might be needed. In a study by Becerril et al. *(33)*, the valence dependency of the display system on the outcome of cell selections was assessed. Phage expressing multivalent scFv (encoded by a phage vector) was more efficiently endocytosed compared to monovalently displayed scFv (encoded by a phagemid vector), thus corroborating previous studies that reported the need for di- or oligomerization of cell surface receptors for efficient endocytosis and receptor signaling *(34–36)*. Affinity was also shown to be of importance, as higher affinity mediated better internalization ratios *(33)*. Nevertheless, the monovalent display of scFv has been proven to give rise to internalizing antibodies *(17,18)*. This might reflect that the isolated antibodies are of high affinity or that they mimic the natural ligand of the targeted receptor. Additionally, increased display levels could have led to more than 1 scFv/phage. Thus, it is likely that the display system used will affect the type of targets that will be identified by specific clones.

2. The cell surface protein composition of the subtractor cell line should optimally be identical to that of the target cell line, with the exemption of the targeted antigen, which should be in excess on the target cell line. To ensure efficient reduction of background and thus increase the likelihood of target specific binders during the selections, the amount of subtractor cells should be in excess compared to the target cells (10–100 times).

3. By far, the most common approach to identify cell surface antigens is immunoprecipitation (ip) of cell lysates. However, since every receptor possesses its own unique properties, e.g., number of transmembrane domains, plasma membrane association (integral or peripheral), glycosylation pattern, and resistance to some detergents, a generally applicable protocol cannot be established. For every receptor, an optimal protocol rather needs to be worked out empirically, by trial and error. Several factors contribute to the outcome of the antigen identification process, including efficiency of membrane protein solubilization, amount of cells and antibody, and stringency of blocking and washing conditions. The initial step is the choice of what detergent to add to the lysis buffer. A wide array of detergents is available from, e.g., Sigma–Aldrich Co (www.sigma.com) whereas Pierce Biotechnology, Inc (www.piercenet.com) and Calbiochem (www.emdbiosciences.com) provide test kits containing several different detergents. Some of the successfully used detergents in cell surface receptor solubilization include the non-ionic detergents Triton X-100 *(19,37,38)*, n-dodecyl-β-maltoside *(39,40)*, and digitonin *(41)*, and the zwitterionic detergent 3-[(3-cholamidopropyl) dimethylammonio]-1-propanesulfonic acid (CHAPS) *(42–44)*. In the subsequent

      wash steps, lower the detergent concentration from 0.5% to 0.1%, since less pro-
tein will be present in these steps.

4. To minimize the loss of internalized phage infectivity, chloroquine (Sigma) could
be added (at 100 μM final concentration) to the phage/cell suspension before incubation at 37°C. The chloroquine will prevent lysosomal acidification and increase
the recovery of infectious phage *(33)*.

5. In the flow cytometry screening procedure, it is recommended to use phage at
a 1/10 dilution. Include negative control (e.g., helper phage) and, if available,
positive control phages displaying target antigen specificity.

6. In the confocal microscopy experiment, it is recommended to include a negative
control. This could be done by incubating cells with phages at 4°C, which should
minimize internalization and thus only result in cell surface localization. In addition, endocytosis inhibitors could be used to monitor this event. The subcellular
localization could be assessed by co-staining with antibodies that are reactive with
different intracellular compartments. For instance, early endosome can be visualized by an EEA1 antibody, whereas late endosomes can be stained by an antibody
against the mannose-6-phosphate receptor.

7. The amount of antibody to use in the immunoprecipitation has to be determined
empirically. Suggested starting amounts range from 5 to 100 μg. In the precipitation step, also add antibody to the other cell lysate fractions generated during the
sequential centrifugation steps.

8. The MALDI-TOF results should, if possible, be corroborated by carrying out
blocking studies with previously characterized antibodies against the identified
target *(45)*. The preferred format of a blocking antibody is a polyclonal, since such
a formulation will contain antibodies binding to most, if not all, epitopes of the target. As an alternative method for verification of the MALDI identification, a small
interfering RNA (siRNA) could be designed specific for the mRNA transcript of
the identified target molecule. In this way, the target could be knocked-down by
RNA interference (RNAi). The binding of antibody to the silenced target cells
could then be assessed by flow cytometry and compared to cells transfected with
a mismatched siRNA sequence *(22)*.

Acknowledgements

We thank Prof. Mats Ohlin for useful comments on the protocols.

References

1. Johannes, L. and C. Lamaze (2002). Clathrin-dependent or not: is it still the ques-
tion? *Traffic* **3**(7): 443–451.
2. Mousavi, S. A., L. Malerod, T. Berg and R. Kjeken (2004). Clathrin-dependent
endocytosis. *Biochem J* **377**(Pt 1): 1–16.
3. Nielsen, U. B., D. B. Kirpotin, E. M. Pickering, K. Hong, J. W. Park, M. Refaat
Shalaby, Y. Shao, C. C. Benz and J. D. Marks (2002). Therapeutic efficacy of

anti-ErbB2 immunoliposomes targeted by a phage antibody selected for cellular endocytosis. *Biochim Biophys Acta* **1591**(1–3): 109–118.

4. Smith, G. P. (1985). Filamentous fusion phage: novel expression vectors that display cloned antigens on the virion surface. *Science* **228**(4705): 1315–1317.

5. Bradbury, A. R. and J. D. Marks (2004). Antibodies from phage antibody libraries. *J Immunol Methods* **290**(1–2): 29–49.

6. Hoogenboom, H. R. (2005). Selecting and screening recombinant antibody libraries. *Nat Biotechnol* **23**(9): 1105–1116.

7. Goletz, S., P. A. Christensen, P. Kristensen, D. Blohm, I. Tomlinson, G. Winter and U. Karsten (2002). Selection of large diversities of antiidiotypic antibody fragments by phage display. *J Mol Biol* **315**(5): 1087–1097.

8. Hawkins, R. E., S. J. Russell and G. Winter (1992). Selection of phage antibodies by binding affinity. Mimicking affinity maturation. *J Mol Biol* **226**(3): 889–896.

9. Persson, H., J. Lantto and M. Ohlin (2006). A focused antibody library for improved hapten recognition. *J Mol Biol* **357**(2): 607–620.

10. Siegel, D. L., T. Y. Chang, S. L. Russell and V. Y. Bunya (1997). Isolation of cell surface-specific human monoclonal antibodies using phage display and magnetically-activated cell sorting: applications in immunohematology. *J Immunol Methods* **206**(1–2): 73–85.

11. Mutuberria, R., H. R. Hoogenboom, E. van der Linden, A. P. de Bruine and R. C. Roovers (1999). Model systems to study the parameters determining the success of phage antibody selections on complex antigens. *J Immunol Methods* **231**(1–2): 65–81.

12. Shadidi, M. and M. Sioud (2001). An anti-leukemic single chain Fv antibody selected from a synthetic human phage antibody library. *Biochem Biophys Res Commun* **280**(2): 548–552.

13. Mutuberria, R., S. Satijn, A. Huijbers, E. Van Der Linden, H. Lichtenbeld, P. Chames, J. W. Arends and H. R. Hoogenboom (2004). Isolation of human antibodies to tumor-associated endothelial cell markers by *in vitro* human endothelial cell selection with phage display libraries. *J Immunol Methods* **287**(1–2): 31–47.

14. Popkov, M., C. Rader and C. F. Barbas, 3rd (2004). Isolation of human prostate cancer cell reactive antibodies using phage display technology. *J Immunol Methods* **291**(1–2): 137–151.

15. Volkel, T., R. Muller and R. E. Kontermann (2004). Isolation of endothelial cell-specific human antibodies from a novel fully synthetic scFv library. *Biochem Biophys Res Commun* **317**(2): 515–521.

16. Huls, G. A., I. A. F. M. Heijnen, M. E. Cuomo, J. C. Koningsberger, L. Wiegman, E. Boel, A.-R. van der Vuurst de Vries, S. A. J. Loyson, W. Helfrich, G. P. van Berge Henegouwen, M. van Meijer, J. de Kruif, et al. (1999). A recombinant, fully human monoclonal antibody with antitumor activity constructed from phage-displayed antibody fragments. *Nat Biotech* **17**(3): 276–281.

17. Poul, M. A., B. Becerril, U. B. Nielsen, P. Morisson and J. D. Marks (2000). Selection of tumor-specific internalizing human antibodies from phage libraries. *J Mol Biol* **301**(5): 1149–1161.

18. Heitner, T., A. Moor, J. L. Garrison, C. Marks, T. Hasan and J. D. Marks (2001). Selection of cell binding and internalizing epidermal growth factor receptor antibodies from a phage display library. *J Immunol Methods* **248**(1–2): 17–30.

19. Gao, C., S. Mao and F. Ronca (2003). De novo identification of tumor-specific internalizing human antibody-receptor pairs by phage-display methods. *J Immunol Methods* **274**(1–2): 185–198.

20. Fransson, J., S. Ek, P. Ellmark, E. Soderlind, C. A. Borrebaeck and C. Furebring (2004). Profiling of internalizing tumor-associated antigens on breast and pancreatic cancer cells by reversed genomics. *Cancer Lett* **208**(2): 235–242.

21. Liu, B., F. Conrad, M. R. Cooperberg, D. B. Kirpotin and J. D. Marks (2004). Mapping tumor epitope space by direct selection of single-chain Fv antibody libraries on prostate cancer cells. *Cancer Res* **64**(2): 704–710.

22. Fransson, J. and C. A. Borrebaeck (2006). The nuclear DNA repair protein Ku70/80 is a tumor-associated antigen displaying rapid receptor mediated endocytosis. *Int J Cancer* **119**(10): 2492–2496.

23. Persson, J. and M. Ohlin (2006). Antigens for the selection of pan-variable number of tandem repeats motif-specific human antibodies against Mucin-1. *J Immunol Methods* **316**(1–2): 116–124.

24. Norderhaug, L., T. Olafsen, T. E. Michaelsen and I. Sandlie (1997). Versatile vectors for transient and stable expression of recombinant antibody molecules in mammalian cells. *J Immunol Methods* **204**(1): 77–87.

25. Griffiths, A. D., S. C. Williams, O. Hartley, I. M. Tomlinson, P. Waterhouse, W. L. Crosby, R. E. Kontermann, P. T. Jones, N. M. Low, T. J. Allison, et al. (1994). Isolation of high affinity human antibodies directly from large synthetic repertoires. *EMBO J* **13**(14): 3245–3260.

26. Knappik, A., L. Ge, A. Honegger, P. Pack, M. Fischer, G. Wellnhofer, A. Hoess, J. Wolle, A. Pluckthun and B. Virnekas (2000). Fully synthetic human combinatorial antibody libraries (HuCAL) based on modular consensus frameworks and CDRs randomized with trinucleotides. *J Mol Biol* **296**(1): 57–86.

27. Söderlind, E., L. Strandberg, P. Jirholt, N. Kobayashi, V. Alexeiva, A. M. Åberg, A. Nilsson, B. Jansson, M. Ohlin, C. Wingren, L. Danielsson, R. Carlsson, et al. (2000). Recombining germline-derived CDR sequences for creating diverse single-framework antibody libraries. *Nat Biotechnol* **18**(8): 852–856.

28. Sidhu, S. S., B. Li, Y. Chen, F. A. Fellouse, C. Eigenbrot and G. Fuh (2004). Phage-displayed antibody libraries of synthetic heavy chain complementarity determining regions. *J Mol Biol* **338**(2): 299–310.

29. Hoet, R. M., E. H. Cohen, R. B. Kent, K. Rookey, S. Schoonbroodt, S. Hogan, L. Rem, N. Frans, M. Daukandt, H. Pieters, R. van Hegelsom, N. C. Neer, et al. (2005). Generation of high-affinity human antibodies by combining donor-derived and synthetic complementarity-determining-region diversity. *Nat Biotechnol* **23**(3): 344–348.

30. McCafferty, J., A. D. Griffiths, G. Winter and D. J. Chiswell (1990). Phage antibodies: filamentous phage displaying antibody variable domains. *Nature* **348**(6301): 552–554.

31. Barbas, C. F., 3rd, A. S. Kang, R. A. Lerner and S. J. Benkovic (1991). Assembly of combinatorial antibody libraries on phage surfaces: the gene III site. *Proc Natl Acad Sci U S A* **88**(18): 7978–7982.

32. O'Connell, D., B. Becerril, A. Roy-Burman, M. Daws and J. D. Marks (2002). Phage versus phagemid libraries for generation of human monoclonal antibodies. *J Mol Biol* **321**(1): 49–56.

33. Becerril, B., M. A. Poul and J. D. Marks (1999). Toward Selection of Internalizing Antibodies from Phage Libraries. *Biochem Biophys Res Commun* **255**(2): 386–393.

34. Yarden, Y. (1990). Agonistic antibodies stimulate the kinase encoded by the neu protooncogene in living cells but the oncogenic mutant is constitutively active. *Proc Natl Acad Sci U S A* **87**(7): 2569–2573.

35. Heldin, C. H. (1995). Dimerization of cell surface receptors in signal transduction. *Cell* **80**(2): 213–223.

36. Hurwitz, E., I. Stancovski, M. Sela and Y. Yarden (1995). Suppression and promotion of tumor growth by monoclonal antibodies to ErbB-2 differentially correlate with cellular uptake. *Proc Natl Acad Sci U S A* **92**(8): 3353–3357.

37. Finbloom, D. S., L. M. Wahl and K. D. Winestock (1991). The receptor for interferon-gamma on human peripheral blood monocytes consists of multiple distinct subunits. *J Biol Chem* **266**(33): 22545–22548.

38. Exley, M., J. Garcia, S. B. Wilson, F. Spada, D. Gerdes, S. M. Tahir, K. T. Patton, R. S. Blumberg, S. Porcelli, A. Chott and S. P. Balk (2000). CD1d structure and regulation on human thymocytes, peripheral blood T cells, B cells and monocytes. *Immunology* **100**(1): 37–47.

39. Peterson, J. A., J. R. Couto, M. R. Taylor and R. L. Ceriani (1995). Selection of tumor-specific epitopes on target antigens for radioimmunotherapy of breast cancer. *Cancer Res* **55**(23 Suppl): 5847s–5851s.

40. Staudinger, R. and J. C. Bandres (2000). Solubilization of the chemokine receptor CXCR4. *Biochem Biophys Res Commun* **274**(1): 153–156.

41. Meenagh, S. A., C. T. Elliott, R. K. Buick, C. A. Izeboud and R. F. Witkamp (2001). The preparation, solubilisation and binding characteristics of a beta 2-adrenoceptor isolated from transfected Chinese hamster cells. *Analyst* **126**(4): 491–494.

42. Naldini, L., D. Cirillo, T. W. Moody, P. M. Comoglio, J. Schlessinger and R. Kris (1990). Solubilization of the receptor for the neuropeptide gastrin-releasing peptide (bombesin) with functional ligand binding properties. *Biochemistry* **29**(21): 5153–5160.

43. Torigoe, K., S. Ushio, T. Okura, S. Kobayashi, M. Taniai, T. Kunikata, T. Murakami, O. Sanou, H. Kojima, M. Fujii, T. Ohta, M. Ikeda, et al. (1997). Purification and characterization of the human interleukin-18 receptor. *J Biol Chem* **272**(41): 25737–25742.

44. Harvey, V., J. Jones, A. Misra, A. R. Knight and K. Quirk (2001). Solubilisation and immunoprecipitation of rat striatal adenosine A(2A) receptors. *Eur J Pharmacol* **431**(2): 171–177.

45. Fransson, J., U. C. Tornberg, C. A. Borrebaeck, R. Carlsson and B. Frendeus (2006). Rapid induction of apoptosis in B-cell lymphoma by functionally isolated human antibodies. *Int J Cancer* **119**(2): 349–358.

9

Artificial Membrane Models for the Study of Macromolecular Delivery

Lena Mäler and Astrid Gräslund

Summary

Artificial biomembrane mimetic model systems are used to characterize peptide–membrane interactions using a wide range of methods. Herein, we present the use of selected membrane model systems to investigate peptide–membrane interactions. We describe methods for the preparation of various membrane mimetic media. Our applications will focus on small unilamellar vesicles (SUVs) and large unilamellar vesicles (LUVs) as well as on media more suited for nuclear magnetic resonance (NMR) techniques, micelles, and fast-tumbling two-component bilayered micelles (bicelles).

Key Words: Phospholipid; Vesicle; Detergent; Micelle; Bicelle.

1. Introduction

A typical biological membrane is a complex structure composed primarily of lipids and proteins. The major structural components of the bilayer are various lipids. In eukaryotes, the most common type of lipids are phosphatidylcholines, whereas in prokaryotes (such as *Escherichia coli*), the main lipids are typically phosphatidylethanolamines *(1)*. One example of a typical eukaryotic neutral (zwitterionic) phospholipid is palmitoyl-oleoyl-phosphatidylcholine (POPC). The molecular structure of POPC is compared to those of dimyristoylphosphatidylcholine (DMPC) and the negatively charged dimyristoylphosphatidylglycerol (DMPG), commonly used in membrane mimetics, in **Fig. 1**.

In order for studies with model membranes to be biologically relevant, it is important that the fluid phase, the lamellar liquid crystalline phase (L_α), is

From: *Methods in Molecular Biology, vol. 480: Macromolecular Drug Delivery,* Edited by: M. Belting
DOI 10.1007/978-1-59745-429-2_9, © Humana Press, a part of Springer Science+Business Media, LLC 2009

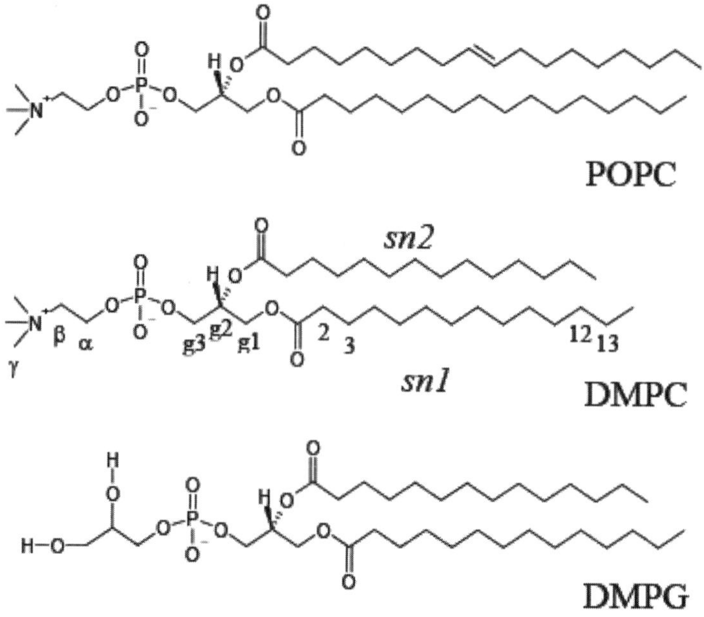

Fig. 1. Molecular structures of POPC, DMPC, and DMPG. The nomenclature for the positioning of the two fatty acyl chains is indicated for DMPC, as is the conventional numbering and labeling of some of the carbon atoms.

predominant. Typical values for the various properties of fluid-phase vesicles containing DMPC are documented in **Table 1**. Other features of a model membrane that need to be considered are bilayer thickness, headgroup size, and headgroup charge. These properties can also, in certain instances, be finely tuned in model membrane systems, e.g., vesicles.

A simple membrane model that, from many perspectives, is most similar to natural biomembranes involves vesicles, sometimes also referred to as liposomes (**Fig. 2**). Vesicles are water-filled spheres delineated by lipid bilayers. It is possible to produce vesicles using different lipid properties, such as headgroup charge and bilayer thickness. A typical biologically relevant vesicle is unilamellar, meaning that only a single bilayer structure is present in each vesicle. Vesicles are classified as small unilamellar vesicles (SUVs), large unilamellar vesicles (LUVs), and giant unilamellar vesicles (GUVs). The peptide-induced membrane perturbations or peptide translocation can be conveniently investigated using unilamellar vesicles. SUVs have been successfully used to investigate the structure and membrane interactions of biologically active peptides, such as the cell-penetrating peptides, by circular dichroism and fluorescence methods (2–4) as well as the peptide-induced leakage (5). LUVs have a larger inner volume

Table 1
Typical Physical Properties of DMPC-Containing Vesicles in the Fluid Phase

Property	Value	Reference
Membrane thickness[a]	44.2 Å	*(37)*
Thickness of the fatty acyl core	26.2 Å	*(37)*
DMPC volume	1100 Å^3	*(37)*
Rate of lateral diffusion	$10^{-11}\,\mathrm{m}^2/\mathrm{s}$	*(38)*
Gel phase/fluid phase transition temperature (L$_\beta$→L$_\alpha$)	298 K	*(39)*
$D_\perp$[b]	$3.3 \times 10^7\,\mathrm{s}^{-1}$	*(40)*
$D_\parallel$[b]	$3.3 \times 10^8\,\mathrm{s}^{-1}$	*(40)*
S_{lipid}[c]	0.58	*(41)*

[a]Steric thickness, without hydration;
[b]$D_\parallel$ and $D_\perp$ are the overall rotational diffusion coefficients, around and perpendicular to the main axis, respectively;
[c]S_{lipid} is the order parameter for the overall lipid motion.

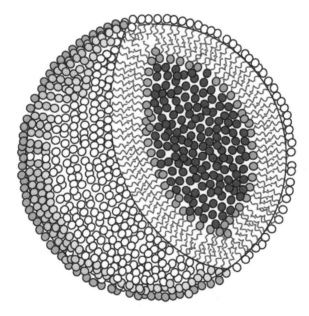

Fig. 2. Cartoon of a phospholipid vesicle. The dark parts indicate the interior of the closed vesicle, and lighter color the outside.

(100 nm in diameter) and lack the curvature stress causing tension, and because of this they are in many ways better as membrane mimetic than the SUVs.

As vesicles are true lipid compartments, it is possible to investigate the processes involving species entering or exiting the vesicles. These studies include peptide-induced leakage, which can be monitored by different methods *(6)* as well as studies on the translocation of peptides across the membranes *(7)*. Furthermore, one can produce vesicles with an uneven distribution of certain lipids in the inner and outer leaflets, which can be used to mimic the real membrane. Vesicles with a pH gradient or electric potential across the bilayer have also been used to investigate potential-mediated effects on peptide translocation and leakage *(8–12)*.

Other vesicular structures can also be used to characterize peptide–membrane interactions. Macroscopic vesicles, GUVs, with a diameter of more than 1 μm can be prepared and these can be studied using ordinary light microscope *(13)*.

Certain spectroscopic techniques, such as nuclear magnetic resonance (NMR) methods, require that the membrane mimetic, i.e., the lipid aggregate is not too large, and that the lipids exhibit a high degree of motion. For such studies, the micellar membrane model is often preferred. Micelles are relatively small (**Fig. 3**, *top*), which means that they rotate rapidly, on the time-scale required for NMR. These micelles consist of detergent molecules that aggregate above the critical micelle concentration (CMC). The size of a micelle is defined by the

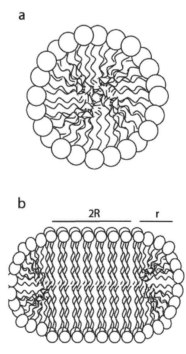

Fig. 3. Schematic view of the cross-section of a micelle (**a**), and a detergent–lipid bicelle (**b**). In (**b**), the diameter of the bilayered region is given as $2R$, while the radius of the detergent is r.

aggregation number, i.e., the average number of detergent molecules present. An increase in the detergent concentration above the CMC results in the formation of more micelles, rather than larger micelles. Examples of detergents that form micelles useful for spectroscopic studies are dihexanoyl phosphatidylcholine (DHPC), sodium dodecyl sulfate (SDS), and dodecyl phosphocholine (DPC).

Micelles of various kinds have proven to be useful as mimetics for investigating the high-resolution structure of bound peptides. These include studies of several cell-penetrating peptides *(14,15)* as well as membrane-disrupting peptides including antimicrobial peptides *(16)*.

Compared to phospholipid vesicles, it has been observed that peptides tend to be structured in micelles, and that SDS in particular seems to induce a high degree of α-helical secondary structure. In other cases, it has been shown that micelle aggregates induce curved helical structures in peptides, which are not seen in more realistic bilayered membrane models *(17,18)*.

The two major drawbacks of micelles as membrane-mimicking models are that they are composed of detergents, and exhibit a strong curvature, which, as already mentioned, may lead to peptides adopting non-native structures. A recent improvement in this connection was the discovery of bicelles *(19,20)*, i.e., disk-shaped aggregates formed by mixing certain lipids and detergents at specific ratios (**Fig. 3**, *bottom*). The properties of such bicelles are dependent on several physical parameters, but one key feature is the q-value, which is defined as the molar ratio between lipid and detergent ($q = $ [lipids]/[detergent]). Larger values of q give rise to aggregates with a more pronounced disk-like shape. A sufficiently high q-value yields a lipid phase that aligns in magnetic fields. The phase behavior of bicelles with large q-values has been found to be highly multifaceted, and conflicting theories have been presented as to whether a bicellar solution consists of discs, or if a completely different phase is formed *(21–26)*. At low q-values, however, bicelles are isotropically tumbling objects, with more or less well-defined size and shape. These bicelles have been used successfully to investigate the details in membrane–peptide interactions *(27,28)*. Several experimental methods, including NMR spectroscopy, electron microscopy, fluorescence spectroscopy, dynamic light scattering, and small-angle neutron scattering, indicate that disk-shaped objects are formed *(29,30)*.

The use of bicelles has been reported to have several advantages compared to micelles. For instance, the integral membrane protein diacylglycerol kinase retains its activity in bicelles, in contrast to micelles *(31)*. Bicelles also have been reported to form stable crystals with the membrane protein bacteriorhodopsin, allowing X-ray crystallographic studies *(32)*. In addition, the interaction of the HIV-1 envelope peptide with micelles induces a strong curvature in this model membrane, which is not observed in the case of bicelles *(17)*.

Nowadays, several solution structures of membrane-bound peptides and protein fragments have been reported *(33,34)*. More important is that bicelles have been used successfully to investigate the effect of bioactive peptides on lipid properties, such as order and dynamics, and at the same time of the dynamics of the peptides themselves *(35,36)*.

2. Materials

2.1. Vesicle Preparation

1. Phospholipids, typically POPC and palmityl-oleoyl-phosphatidylglycerol (POPG); Avanti Polar Lipids (Alabaster, AL, USA)
2. Phosphate buffer, pH 7 (typically 50 mM)
3. Sonicator
4. Lyophilizer
5. Polycarbonate filter with 100 nm pores; Avanti Polar Lipids (Alabaster, AL, USA)

2.2. Micelle Preparation

1. Detergents, typically one of the following: sodium dodecyl phosphate (SDS); Sigma (St. Louis, USA), DPC, 1,2-dihexanoyl-*sn*-glycero-3-phosphocholine (DHPC); Avanti Polar Lipids, Alabaster (AL, USA). For high-resolution NMR studies, deuterated versions of the detergents are often used.
2. Phosphate buffer, pH 5.5–7 (typically 50 mM).

2.3. Preparation of Fast-Tumbling Bicelles (Mixed Bilayered Micelles)

1. Deuterated phospholipids 1,2-dimyristoyl-*sn*-glycero-3-phosphocholine (DMPC-d_{54}), 1,2-dimyristoyl-*sn*-glycero-3-[phospho-*rac*-(1-glycerol)] (DMPG-d_{54}), and 1,2-dihexanoyl-*sn*-glycero-3-phosphocholine (DHPC-d_{22}) as well as the undeuterated DMPC, DMPG, and DHPC; Avanti Polar Lipids, Alabaster (AL, USA)
2. Phosphate buffer, pH 5.5. Usually a 0.2M stock solution is prepared from which enough is taken for preparing the desired concentration (50 mM)
3. Lyophilizer
4. Vortex
5. Oven

3. Methods

3.1. Small Unilamellar Vesicles

1. **Preparing lipid suspension**. Mix the desired lipids in an aqueous medium, typically 50 mM phosphate buffer, and vortex the mixture. This procedure results in a dispersion of multilamellar vesicles (MLVs), which are onion-like bilayer structures containing several bilayers separated by water.

2. **Sonication**. Sonicate the dispension for a short period of time (typically tens of minutes), which results in a solution of SUVs with a diameter within the range of 20–50 nm.
3. **Ultracentrifugation**. As this procedure results in a distribution of vesicle sizes, ultracentrifugation may be used to ensure a uniform size of the vesicles (*see* **Note 2**).

3.2. Large Unilamellar Vesicles

The total lipid concentration in vesicle suspensions is typically in the range $10\mu M$–1 mM, and the amounts of lipids and solvent are determined accordingly.

1. **Preparing lipid suspension**. Dissolve the desired lipids in an organic solvent (chloroform).
2. Lyophilize the solution to produce a dry film of lipids (*see* **Note 1**).
3. Dissolve the dry lipids in a suitable buffer (50 mM phosphate buffer) to give a suspension of large MLVs.
4. **Preparing LUVs from lipid suspension**. The solution is subjected to several (typically four) freeze–thaw cycles in order to decrease the lamellarity.
5. To obtain LUVs with a narrow size distribution, the solution is extruded several times (typically 20 times) through a polycarbonate filter. This results in LUVs with a diameter of 100 nm *(42)* (*see* **Note 3**).

3.2.1. Preparation of Fast-Tumbling Bicelles (Mixed Bilayered Micelles)

For high-resolution solution NMR work on peptides or protein fragments in bicelle solution, chain-deuterated lipids (DMPC, DMPG, and DHPC) are frequently used. The final concentration of lipids and detergents should be between 150 and 300 mM. Typically, a total concentration of 300 mM is used. The ratio of lipid/detergent (the q-value) suitable for high-resolution NMR studies is within the range of 0.15–0.5. For a sample of $q = 0.5$ DMPC/DHPC bicelles, this implies a final DMPC concentration of 100 mM and a DHPC concentration of 200 mM. Surface-active peptides or protein fragments may be added to ready-made bicelle solution.

1. **Preparing lipid suspension**. Mix the desired lipids in an aqueous medium, and vortex the mixture until a homogeneous slurry is obtained. Typically, chain-deuterated lipids (and detergents) are used for NMR purposes (*see* **Note 5**).
2. **Preparation of DHPC stock solution**. Prepare a stock solution of the detergent in water (or D$_2$O if desired for NMR purposes). A 1M solution of DHPC is convenient for preparing bicelles with a total lipid (phospholipids + detergent) concentration of 300 mM (*see* **Note 6**).
3. **Bicelle preparation**. Add an appropriate amount of the detergent solution to the lipid slurry to produce the desired q-value (*see* **Note 7**).

4. The sample is vortexed and heated to well above the melting temperature of the lipid (typically heated to 45°C for DMPC/DHPC mixtures) in several cycles, until a transparent low-viscous solution is obtained (*see* **Notes 8–10**).

3.3. Reconstituting Hydrophobic Peptides in Bicelles

For hydrophobic peptides and protein fragments (such as transmembrane segments), it is advisable to reconstitute them into the bicelle bilayer by adding them to the lipids in an organic solvent prior to preparing the bicelles according to the description below.

1. **Preparing lipid/bicelle solution**. Dissolve the desired lipids in an organic solvent (trifluoroethanol, TFE) together with the peptide.
2. Lyophilize the solution to produce a dry film of lipids.
3. **Preparing the bicelles**. Dissolve the dry lipids in a suitable buffer (50 mM phosphate buffer) and enough stock solution of the detergent to produce the desired q-value and total lipid concentration, as described above.
4. If necessary, lyophilize once more to remove traces of TFE (or other organic solvents), and dissolve in water again.
5. The sample is vortexed and heated to well above the melting temperature of the lipid (typically heated to 45°C for DMPC/DHPC mixtures) in several cycles, until a transparent low-viscous solution is obtained.

4. Notes

1. For the preparation of LUVs, it is important to completely remove all organic solvents initially used for solvation, before buffer is added to make the lipid suspension. Insufficient lyophilization will give rise to incorrect results.
2. Small vesicles are metastable structures, and are only stable for a limited time, typically a few days.
3. Large vesicles can be stored up to months provided an inert atmosphere is provided.
4. It is valuable, if you can check on the final result of your vesicle preparation by doing a dynamic light scattering measurement. This would report on the homogeneity of the preparation and an average vesicle size.
5. For preparation of bicelles, it is essential that the lipids are properly suspended in buffer solution prior to the addition of the DHPC solution. If this is the case, a clear low-viscous solution is formed.
6. It is convenient to use a DHPC stock solution for bicelle preparation. As DHPC is extremely hygroscopic, it is otherwise difficult to estimate the true amount of DHPC that is actually added to the lipid suspension. Furthermore, unless a stock solution is prepared, DHPC should be handled in a dry atmosphere. It is important for the estimation of bicelle size (related to the q-value) that the relative concentrations of lipids and DHPC can be controlled.
7. It has been shown that some of the DHPC exists as free monomers in solution, which leads to overestimating the DHPC content in the bicelles, and, hence, to underestimating the true bicelle size (related to the q-value).

8. Bicelles with different q-values can be produced using the methodology outlined. It has, however, been reported that a minimum of total lipid concentration (lipid + detergent) of around 100 mM is required in order to maintain the bicelle's disk-shape *(29)*.
9. Bicelle solutions can be stored in the freezer ($-20°$C) for months. Bicelles made with negatively charged lipids (DMPG) are, however, more sensitive to degradation.
10. Bicelles can be prepared with lipids with varying acyl chain lengths, such as 1,2-dilauroyl-*sn*-glycero-3-phosphocholine (DLPC, 12 C) and dipalmitoyl-*sn*-glycero-3-phosphocholine (DPPC, 16 C). It should, however, be noted that the morphology of such mixtures has not yet been fully characterized.

References

1. Gennis, R.B. (1989) Biomembranes: Molecular Structure and Function. Springer, New York.
2. Magzoub, M., K. Kilk, L.E.G. Erikssom, U. Langel, and A. Gräslund (2001) Interaction and structure induction of cell-penetrating peptides in the presence of phospholipid vesicles. *Biochim. Biophys. Acta* **1512**, 77–89.
3. Magzoub, M., L.E.G. Eriksson, and A. Gräslund (2002) Conformational states of the cell-penetrating peptide penetratin when interacting with phospholipid vesicles: effects of surface charge and peptide concentration. *Biochim. Biophys. Acta* **1563**, 53–63.
4. Magzoub, M., L.E.G. Eriksson, and A. Gräslund (2003) Comparison of the interaction, positioning, structure induction and membrane perturbation of cell-penetrating peptides and non-translocating variants with phospholipid vesicles. *Biophys. Chem.* **103**, 271–288.
5. Weinstein J.N., S. Yoshikami, P. Henkart, R. Blumenthal, and W.A. Hagins (1977) Liposome-cell interaction: transfer and intracellular release of a trapped fluorescent marker. *Science* **195**, 489–492.
6. Andersson, A., J. Danielsson, A. Gräslund, and L. Mäler (2007) Kinetic models for peptide-induced leakage from vesicles and cells. *Eur. Biophys. J.* **36**, 621–635.
7. Magzoub, M., and A. Gräslund (2004) Cell-penetrating peptides: from inception to application. *Q. Rev. Biophys.* **37**, 147–195.
8. Terrone, D., S. Leung, W. Sang, L. Roudaia, and J. Silvius (2003) Penetratin and related cell-penetrating cationic peptides can translocate across lipid bilayers in the presence of a transbilayer potential. *Biochemistry* **42**, 13787–13799.
9. Thorén, P., D. Persson, E. Esbjörner, M. Goksör, B. Lincoln, and B. Nordén (2004) Membrane binding and translocation of cell-penetrating peptides. *Biochemistry* **43**, 3471–3489.
10. Bárány-Wallje, E., S. Keller, S. Serowy, S. Geibel, P. Pohl, M. Bienert, and M. Dathe (2005) A critical reassessment of penetratin translocation across lipid membranes. *Biophys. J.* **89**, 2513–2521.
11. Magzoub, M., A. Pramanik, and A. Gräslund (2005) Modeling the endosomal escape of cell-penetrating peptides: transmembrane pH gradient driven translocation across phospholipid bilayers. *Biochemistry* **44**, 14890–14897.
12. Björklund, J., H. Biverståhl, A. Gräslund, L. Mäler, and P. Brzezinski (2006) Real.time transmembrane translocation of penetratin driven by light-generated proton pumping. *Biophys. J. Biophys. Lett.* **91**, L29–L31.

13. Fischer, A., T. Oberholzer, and P.L. Luisi (2000) Giant vesicles as models to study the interactions between membranes and proteins. *Biochim. Biophys. Acta* **1467**, 177–188.

14. Lindberg, M., J. Jarvet, U. Langel, and A. Gräslund (2001) Secondary structure and position of the cell-penetrating peptide transportan in SDS micelles as determined by NMR. *Biochemistry* **40**, 3141–3149.

15. Biverståhl, H., A. Andersson, A. Gräslund and L. Mäler (2004) NMR solution structure and membrane interaction of the N-terminal sequence (1–30) of the bovine prion protein. *Biochemistry* **43**, 14940–14947.

16. Damberg, P., J. Jarvet, and A. Gräslund (2001) Micellar systems as solvents in peptide and protein structure determination. *Meth. Enzymol.* **339**, 271–285.

17. Chou, J.J., J.D. Kaufman, S.J. Stahl, P.T. Wingfield, and A. Bax (2002) Micelle-induced curvature in a water-insoluble HIV-1 env peptide revealed by NMR dipolar coupling measurement in stretched polyacrylamide gel. *J. Am. Chem. Soc.* **124**, 2450–2451.

18. Andersson A., and L. Mäler (2002) NMR solution structure and dynamics of motilin in isotropic phospholipid bicellar solution. *J. Biomol. NMR* **24**, 103–112.

19. Ram, P., and J.H. Prestegard (1988) Magnetic field induced ordering of bile salt/phospholipid micelles: new media for NMR structural investigations. *Biochim. Biophys. Acta* **940**, 289–294.

20. Sanders, C.R., and J.H. Prestegard (1990) Magnetically orientable phospholipid bilayers containing small amounts of a bile salt analogue, CHAPSO. *Biophys. J.* **58**, 447–460.

21. Gaemers, S., and A. Bax (2001) Morphology of three lyotropic liquid crystalline biological NMR media studied by translational diffusion anisotropy. *J. Am. Chem. Soc.* **123**, 12343–11235.

22. Arnold, A., T. Labrot, R. Oda, and E.J. Dufourc (2002) Cation modulation of bicelle size and magnetic alignment as revealed by solid-state NMR and electron microscopy. *Biophys. J.* **83**, 2667–2680.

23. Nieh, M.P., V.A. Raghunathan, C.J. Glinka, T.A. Harroun, G. Pabst, and J. Katsaras (2004) Magnetically alignable phase of phospholipid "bicelle" mixtures is a chiral nematic made up of wormlike micelles. *Langmuir* **20**, 7893–7897.

24. van Dam, L., G. Karlsson, and K. Edwards (2004) Direct observation and characterization of DMPC/DHPC aggregates under conditions relevant for biological solution NMR. *Biochim. Biophys. Acta* **1664**, 241–256.

25. Triba, M.N., D.E. Warschawski, and P.F. Devaux (2005) Reinvestigation by phosphorus NMR of lipid distribution in bicelles. *Biophys. J.* **88**, 1887–1901.

26. Triba, M.N., P.F. Devaux, and D.E. Warschawski (2006) Effects of lipid chain length and unsaturation on bicelles stability. A phosphorus NMR study. *Biophys. J.* **91**, 1357–1367.

27. Vold, R.R., and R.S. Prosser (1996) Magnetically oriented phospholipid bilayer micelles for structural studies of polypeptides. Does the ideal bicelle exist? *J. Magnet. Reson.* **113**, 267–271.

28. Vold, R.R., S.R. Prosser, and A.J. Deese (1997) Isotropic solutions of phospholipid bicelles: a new membrane mimetic for high-resolution NMR studies of polypeptides. *J. Biomol. NMR* **9**, 329–335.

29. Glover, K.J., J.A. Whiles, G. Wu, N.-J. Yu, R. Deems, J.O. Struppe, R.E. Stark, E.A. Komives, and R.R. Vold (2001) Structural evaluation of phospholipid bicelles

for solution-state studies of membrane-associated biomolecules. *Biophys. J.* **81**, 2163–2171.

30. Luchette, P.A., T.N. Vetman, R.S. Prosser, R.E.W. Hancock, M.P. Nieh, C.J. Glinka, S. Krueger, and J. Katsaras (2001) Morphology of fast-tumbling bicelles: a small angle neutron scattering and NMR study. *Biochim. Biophys. Acta* **1513**, 83–94.

31. Sanders, C.R.I., and G.C. Landis (1995) Reconstitution of membrane proteins into lipid-rich bilayered mixed micelles for NMR studies. *Biochemistry* **34**, 4030–4040.

32. Faham, S., and J.U. Bowie (2002) Bicelle crystallization: a new method for crystallizing membrane proteins yields a monomeric bacteriorhodopsin structure. *J. Mol. Biol.* **316**, 1–6.

33. Marcotte, I., and M. Auger (2005) Bicelles as model membranes for solid- and solution-state NMR studies of membrane peptides and proteins. *Concepts Magnet. Reson.* **24**, 17–37.

34. Prosser, R.S., F. Evanics, J.L. Kitevski, M.S. Al-Abdul-Wahid (2006) Current applications of bicelles in NMR studies of membrane-associated amphiphiles and proteins. *Biochemistry* **45**, 8453–8465.

35. Andersson A., and L. Mäler (2005) Magnetic resonance investigations of lipid motion in isotropic bicelles. *Langmuir* **21**, 7702–7709.

36. Bárány-Wallje E., A. Andersson, A. Gräslund, and L. Mäler (2006) Dynamics of transportan in bicelles is surface charge dependent. *J. Biomol. NMR* **35**, 137–147.

37. Nagle, J.F., and S. Tristram-Nagle (2000) Structure of lipid bilayers. *Biochim. Biophys. Acta* **1469**, 159–195.

38. Orädd, G., and G. Lindblom (2004) NMR Studies of lipid lateral diffusion in the DMPC/gramicidin D/water system: peptide aggregation and obstruction effects. *Biophys. J.* **87**, 980–987.

39. Hinz, H.J., and J.M. Sturtevant (1972) Calorimetric investigation of the influence of cholesterol on the transition properties of bilayers formed from synthetic L-lecithins in aqueous suspension. *J. Biol. Chem.* **247**, 3697–3700.

40. Mayer, C., G. Gröbner, K. Müller, K. Weisz, and G. Kothe (1990) Orientation-dependent deuteron spin-lattice relaxation times in bilayer membranes: characterization of the overall lipid motion. *Chem. Phys. Lett.* **165**, 155–161.

41. Ellena, J.F., L.S. Lepore, and D.S. Cafiso (1993) Estimating lipid lateral diffusion in phospholipid vesicles from ^{13}C spin–spin relaxation. *J. Phys. Chem.* **97**, 2952–2957.

42. Mayer, L.D., M.J. Hope, and P.R. Cullis (1986) Vesicles of variable sizes produced by a rapid extrusion procedure. *Biochim. Biophys. Acta* **858**, 161–168.

10

Enhanced Delivery of Macromolecules into Cells by Electroendocytosis

Alexander Barbul, Yulia Antov, Yosef Rosenberg, and Rafi Korenstein

Summary

Transfer of exogenous material into the cytosol of cells is one of the main challenges in drug delivery. We present a novel physical approach for efficient incorporation of macromolecules into living cells, based on exposing them to a train of unipolar electric field pulses, possessing much lower amplitude than used for electroporation. The exposure of cells to a low electric field (LEF) alters the cell surface, leading to enhanced adsorption of macromolecules and their subsequent uptake by stimulated endocytosis. The macromolecules are initially encapsulated in membrane vesicles and then, at a later stage, are released into the cytosol and interact with intracellular targets. The uptake of fluorescently labeled macromolecules is monitored using confocal microscopy and flow cytometry. The biological activities of the incorporated macromolecules are determined by biochemical methods.

Key Words: Pulsed low electric fields; Endocytosis; Uptake kinetics; Membrane internalization; Albumin; Lucifer Yellow; Propidium iodide; Adsorption; Flow cytometry; Confocal microscopy.

1. Introduction

The introduction of macromolecules such as antibodies, enzymes, and drugs into the cytosolic compartment of cells, without compromising their biological activities, constitutes a challenge for the introduction of novel therapeutic modalities. Thus, for the last several decades, different approaches were singled out to achieve this target, based on a combination of chemical, physical, and biological approaches (1). There are three main types of physical methodologies that lead to simultaneous incorporation of molecules into a population of cells.

From: *Methods in Molecular Biology, vol. 480: Macromolecular Drug Delivery,* Edited by: M. Belting
DOI 10.1007/978-1-59745-429-2_10, © Humana Press, a part of Springer Science+Business Media, LLC 2009

The first method employs the "ballistic gun" *(2,3)*, where cells are exposed to ballistic bombardment by microparticles coated with the molecules of choice (e.g., DNA). The second method is based on exposing the cells to ultrasound leading to an increased transmembrane transport *(4)*. The third approach is based on an electrically driven process (electroporation), where cells are exposed to high-electric fields for short durations of micro- to milliseconds *(5)*. This exposure leads to induction of short-lived permeability changes in the membrane ("pores") enabling the diffusion of molecules across the membrane along their electrochemical gradients.

It should be stressed that, the mechanism of uptake common to all the above-mentioned methods is the formation of artificial transient permeability pathways in the cell membrane leading to the introduction of extracellular molecules into the intracellular milieu. An alternative principal approach to the incorporation of macromolecules into cells may be based on the enhancement of a naturally occurring process, such as endocytosis, which is involved in the uptake of macromolecules *(6–8)*. An early attempt to enhance uptake via endocytic pathways employed osmotic lysis of pinocytic vesicles *(9)*. This approach involves a brief exposure of cells to a hypertonic solution containing the molecule to be incorporated with subsequent addition of a hypotonic media, which lyse the pinosomes formed during the hypertonic treatment. A frequently used endocytic-based method of uptake employs a ligand conjugated to the appropriate molecule, which is introduced into the cell through receptor-mediated endocytosis *(10–12)*. An example of the therapeutic impact of such an approach is the employment of folate-conjugated molecules that are endocytosed through folate receptors, which are overexpressed in human cancers. This methodology is being utilized for the selective delivery of imaging and therapeutic agents to tumor tissue *(13)*.

We have introduced a novel method to enhance the uptake of macromolecules via stimulating endocytic-like processes by exposing cells to a train of pulsed low electric field, LEF *(14–16)*. The enhanced uptake is attributed both to the direct stimulation of different endocytic pathways as well as to the indirect effect mediated through the increase of the adsorption of the macromolecules onto the exposed cells. This method has found an application in the treatment of different metastatic tumor models *(17–20)*.

In this chapter, we present the basic protocols for applying electroendocytosis for enhanced incorporation of macromolecules into the cells.

2. Materials

2.1. Cell Culture

1. Cell culture medium, e.g., Dulbecco's modified Eagle's medium (DMEM) supplemented with L-glutamine (2 mM), 10% fetal calf serum (FCS), and 0.05%

PSN solution (penicillin 10,000 units/mL, streptomycin 10 mg/mL, and nystatin 1250 units/mL) or Minimum Essential Medium Eagle with non-essential amino acids (MEM-NAA) medium supplemented with L-glutamine (2 m*M*), 5% FCS, and 0.05% PSN solution, or other as required for each cell type.
2. Trypsin/ethylenediaminetetraacetic acid (EDTA): 0.05% trypsin, 0.53 m*M* EDTA in phosphate buffered saline (PBS).
3. HEPES 1-*M* sterile solution (Biological Industries, Beit Haemek, Israel).

2.2. Molecular Probes

Dextran conjugated to fluorescein-5-isothiocyanate (dextran-FITC, m.w. 20–2000 kDa, 0.009 mol FITC per mole glucose) is used at a final concentration of 0.2 μ*M*. Bovine serum albumin conjugated to fluorescein-5-isothiocyanate (BSA-FITC) containing 12 mol FITC per mole albumin (m.w. 66 kDa) is applied at 1–10 μ*M* concentration. Dye-conjugated probes are dialyzed before use against the exposure medium. Propidium iodide (PI) is used at a final concentration

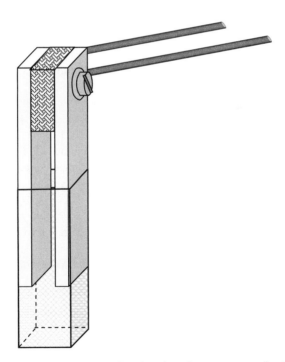

Fig. 1. Schematic representation of a chamber, for exposure of cell suspensions to low electric field. Two stainless steel electrodes separated by a 5-mm Teflon spacer are introduced into a 1-cm plastic cuvette.

of 30 μ*M* for the detection of damaged cells. These markers were purchased from Sigma Chemicals (Rehovot, Israel).

2.3. Equipment

1. Exposure of cells to low intensity trains of unipolar rectangular voltage pulses is carried out by electric pulse generator (e.g., Grass S44 Stimulator, West Warwick, RI, USA).
2. The exposure is performed in a 1-cm plastic cuvette by placing the cell suspension between the two parallel stainless steel or platinum electrodes separated by 0.5 cm, yielding an uniform electric field (**Fig. 1**).
3. Oscilloscope to control voltage and electric pulse parameters.
4. Wide-band current transformer (Pearson Electronics, Palo Alto, CA, USA) to control chamber current.
5. For analysis of the uptake of fluorescent-labeled molecules: flow cytometer or ELISA fluorescent plate reader, confocal fluorescent microscope are used.

3. Methods

3.1. Exposure of Cells to Pulsed Low Electric Fields

3.1.1. Preparation of Cells

1. The cells should be plated at the appropriate density, so that they will be harvested at a subconfluent stage in the log phase of growth.
2. On the day of experiment, trypsinize the adherent cell cultures (*see* **Note 1**). After washing twice by centrifugation (5 min, $210 \times g$), resuspend the cells in the exposure serum-free culture medium, e.g., DMEM supplemented with 25 m*M* HEPES (DMEM-H). Count viable cells and adjust the cell density to a range of 0.5–3×10^6 cells/mL. Maintain the cells at room temperature.

3.1.2. Electroendocytosis

1. Add the macromolecules at the appropriate concentration to the cell suspension (*see* **Note 2**).
2. Place 0.5 mL of the cell suspension containing the macromolecules in the exposure chamber (*see* **Note 3**).
3. Dip electrodes into the cuvette (**Fig. 1**).
4. Apply train of unipolar rectangular pulses with duration of 90–900 μs and frequency of 100–1000 Hz for total time of exposure of 0.5–3 min (*see* **Notes 4 and 5**).
5. Incubate for 4 min at room temperature (*see* **Note 6**).
6. Wash the cells twice by gentle centrifugation for 5 min at $100 \times g$ and resuspend in the growing medium supplemented with serum and PSN.
7. Incubate cells under optimal conditions until analysis (*see* **Note 7**).

3.2. Analysis of Electroendocytosis

3.2.1. Measurement of Adsorption and Uptake by Flow Cytometry, Fluorimetry, and Confocal Fluorescence Microscopy

1. Cells (usually 3×10^6/mL) are exposed to LEF in the presence of the macro-molecular probe of choice (e.g., BSA-FITC at 6.8 μM concentration; *see* **Notes 8 and 9**).

2. After exposure termination and further 4 min of incubation with the probe, cells are washed twice with DMEM-H medium and incubated in growth medium for additional 25 min until analysis.

3. Cell viability is assessed by the PI exclusion test. For flow cytometry and confocal microscopy, PI (30 μM) is added to exposed and control cells 5 min before analysis.

4. Some macromolecular probes (e.g., BSA-FITC) were found to be significantly adsorbed onto the cell membrane, especially after exposure to LEF. In order to differentiate between internalized and adsorbed fractions, the cells are subjected, in the case of proteins, to 0.01% trypsin in PBS for 5 min at 37°C. The trypsin-digested BSA-FITC represents the amount of BSA adsorbed onto the cells, whereas the amount of the probe measured in the cells after trypsinization is attributed to the internalized fraction.

5. The efficiency of removal of the adsorbed BSA-FITC probe from the cell surface by trypsin is validated by confocal fluorescence microscopy (LSM 410, Zeiss, Jena, Germany) using 40 × water immersion objective, 488-nm argon laser excitation, and 500–550 nm band-pass emission filter. To visualize the cellular location of the probe in the exposed cells by confocal microscopy, sequential images of 0.5-μm optical sections are taken (**Fig. 2a,b** and Color Plate 2, *see* Color Plate Section). The red fluorescence of PI, detected via long-pass 585 nm filter using the same excitation wavelength, is used for the detection and elimination of damaged cells from cell population undergoing analysis of uptake.

6. The quantification of uptake is performed by flow cytometry (FACSort, Becton Dickinson, San Jose, CA, USA). Flow cytometry analysis is carried out by employing 488-nm argon laser excitation, and the green fluorescence of FITC is measured via 530/30-nm filter. For the assortment of dead cells, the red fluorescence of PI is detected via 585/42-nm filter. To eliminate signals due to cellular fragments, only those events with forward and side light scattering comparable to whole cells are analyzed. All fluorescence signals are logarithmically displayed. Ten thousands cells are run for each sample and data are collected in the list mode. The analysis of flow cytometry data is performed using WinMDI 2.8 flow cytometry application software. Results are expressed in terms of cell quantities at 95% confidence interval or as the geometric mean of cell population (*see* **Note 10**). The geometric mean (G_m) might be used for log-amplified data, as it takes into account the weighting of data distribution (**Fig. 2c**). Efficiencies of probe adsorption and uptake can be characterized by fold of induction (FI), which is calculated as a ratio of the geometric mean of fluorescence in an exposed sample to that of a control unexposed one.

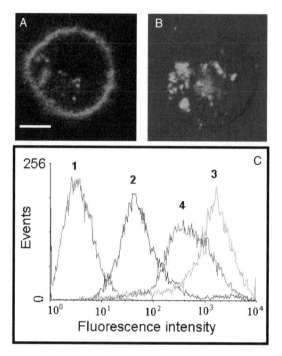

Fig. 2. Low electric field-stimulated adsorption and uptake of BSA-FITC by COS 5–7 cells. Cell suspensions of COS 5–7 cells in DMEM-H supplemented with 6.8 μM BSA-FITC are subjected to LEF treatment of 20 V/cm (180 μs pulse duration at frequency of 500 Hz) for 1 min, followed by two successive washings with DMEM-H solution. Cells are observed by confocal microscopy. **a** and **b** are confocal images of FITC fluorescence taken at central optical section through the *X–Y* plane of COS 5–7 cells, in **b** it is superimposed also with phase contrast image (*blue channel*). Bar = 5 μm. **(a)** Imaging immediately after washing with DMEM-H. It can be seen that a large amount of BSA-FITC is adsorbed and some is already internalized. **(b)** Imaging after 25 min of incubation at 24°C followed by trypsinization. It resulted in an effective uptake of BSA-FITC and its removal from the plasma membrane. **(c)** Distribution histograms of BSA-FITC fluorescence intensity in COS 5–7 cells measured by flow cytometry 1 h after exposure to LEF and incubation at 24°C: *(1)* control cells in the absence of BSA-FITC in the medium; *(2)* control cells in the presence of BSA-FITC – constitutive uptake; *(3)* cells exposed to LEF in the presence of BSA-FITC – enhanced adsorption and uptake; *(4)* same exposed cells as in (3), but further subjected to trypsinization before measurement – enhanced uptake only (*see* Color Plate 2, *see* Color Plate Section).

7. A direct measurement of the amount of BSA-FITC adsorbed onto the cells is carried out by measuring the fluorescence of the probe in the extracellular medium.
   After the release of fragmented BSA-FITC from the cell surface by trypsinization
   (as in **step 4**), cells are sedimented by centrifugation for 5 min at $100 \times g$, and

the supernatant transferred to 96-well flat bottom plates. Cells are resuspended in the same volume of fresh DMEM-H medium and also transferred to the plates. FITC fluorescence is measured using a microplate fluorescence reader (FL-600, Bio-tek, Winooski, VT, USA; excitation filter= 485/20, emission filter=530/25). The readings are normalized to the fraction removed from the unexposed samples. Intracellular uptake can also be estimated by normalization of the fluorescence of wells with LEF treated cells to that of untreated ones, though this way is less precise than flow cytometry as it does not take into account signals from dead cells.

4. Notes

1. Electroendocytic uptake of macromolecules can be induced by LEF treatment of adherent cells as well. There are two ways to apply electric field pulses to cell cultures growing on plastic or glass surfaces: (a) via block of equidistant stainless steel needle electrodes connected in an alternating modes to the (+) and the (−) poles of the pulse generator; (b) by an array of parallel stainless steel electrodes. Method (a) has the advantage of reduced cell damage, but electric field distribution is inhomogeneous possessing maximal values in the immediate vicinity of the needle electrodes. When applying method (b) the produced electric field is homogeneous, but more cells are physically damaged in close proximity of the electrodes. Adherent cells are washed twice with DMEM-H medium, the macromolecule to be introduced is added at the proper concentration, electrodes are applied to the cell culture, and exposure is carried out as described.

2. Small water-soluble molecules can also be introduced into cells via stimulation of fluid-phase uptake by the described protocol. Uptake of Lucifer Yellow (m.w. 457.2; Lucifer Yellow CH, dilithium salt from Sigma Chemicals, Rehovot, Israel) has been stimulated by 2.2-fold and 2.7-fold in COS 5-7 and HaCaT cell lines, respectively, as compared with the constitutive uptake.

3. When performing experiments with cells for a long-time period, following exposure one is required to perform electroendocytosis under sterile conditions. Thus, the electrodes are sterilized with 70% ethanol for 30 min, with subsequent washing with sterile water and drying under laminar flow. Sterile cuvettes (e.g., from Elkay Products, Inc., Shrewsbury, MA, USA) should be used, and the whole procedure of electric treatment and incubation should be performed in a laminar hood.

4. Human keratinocytes HaCaT (gift from Prof. N.E. Fusenig of Deutsches Krebsforschungszentrum, Heidelberg) and HeLa cell lines, African green monkey kidney COS 5–7 (fibroblast-like cells, derived from CV-1 subclone of COS 7), and Chinese Hamster Ovary (CHO) cells have been used with this protocol. The optimal conditions for electroendocytic uptake may vary between different cell lines and should be empirically determined by maximization of uptake and minimization of cell damage. The best parameters of exposure for COS 5–7 and HaCaT cell lines are electric field strength of 20 V/cm, 180 μs pulse duration, frequency of 500 Hz, and total exposure time of 1 min. The electric field parameters can be validated by online monitoring of the voltage and current (e.g., by using a

wide-band current transformer (Pearson Electronics, Palo Alto, CA, USA) on an oscilloscope.

5. The application of a typical train of pulses, when using stainless steel electrodes, resulted in small polarization of the electrodes and the appearance of a residual low DC component (≤ 2 V). To avoid polarization, it is recommended to switch electrode polarity after each treatment and to clean and polish electrodes after each experiment. When employing stainless steel electrodes one observes, in some culture media, the formation of precipitates. It is, therefore, recommended to employ platinum electrodes, which do not lead to precipitation.

6. The electric field-stimulated uptake of macromolecules is temperature dependent whereas stimulated adsorption is not. For example, the uptake of BSA-FITC by COS 5–7 cells is almost completely diminished at 4°C and enhanced twofold at 37°C compared with room temperature. When elevating the temperature during exposure, one should take into account that LEF treatment also leads to temperature elevation in the cell suspension. The temperature of the solutions during exposure can be measured using fiber-optic temperature sensors (FISO Technologies, Quebec, Canada). Transient temperature rise by up to 2°C can be measured at the end of 1 min exposure of DMEM-H medium to LEF (20 V/cm, 180 μs pulse duration, and 500 Hz frequency).

7. Macromolecules initially enter the cells inside membrane vesicles (endosomes). Release of the macromolecules from endocytic vesicles to the cytosol (endosomal escape) is a continuous process that takes a few hours. For example, activity of alkaline phosphatase toward intracellular targets could be detected only 1–3 h after the electroendocytic uptake (*see* **Note 8**).

8. Electric field-stimulated endocytosis can be applied to enhance uptake of biologically active macromolecules into the cells (*see also* **Note 9**). LEF induces uptake of alkaline phosphatase into cells which leads to non-selective change in the phosphorylation state of intracellular proteins. Alkaline phosphatase at 10 μg/mL or 50 μg/mL is added to a suspension of COS 5–7 cells in DMEM-H medium. Following 5 min of preincubation, the cells are subjected to 1 min exposure to LEF (20 V/cm, 180 μs pulse duration, and frequency of 500 Hz). The level of epidermal growth factor receptor (EGFR) phosphorylation is determined by Western blot analysis employing anti-phosphotyrosine and anti-EGFR antibodies. High dephosphorylation of EGFR was found in cells exposed to LEF, when alkaline phosphatase was employed at 50 μg/mL resulting in a 73% decrease in receptor phosphorylation compared to cells exposed to LEF in the absence of alkaline phosphatase in the external medium. The employment of a lower concentration of alkaline phosphatase of 10 μg/mL decreased the phosphorylation of EGFR by 60% compared to controls. Maximal effect of EGFR dephosphorylation was detected at 3 h after alkaline phosphatase loading.

9. The ability of LEF to induce incorporation of macromolecules into the cytosol of living cells, while maintaining their biological activity, can be demonstrated by monitoring activity of horseradish peroxidase (HRP). The activity is determined 3 h following the exposure, after a relatively long time which allows for advanced processing of the incorporated enzyme. COS 5–7 and HaCaT cells

COS 5-7

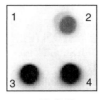
HaCaT

Fig. 3. Activity of HRP following LEF induced uptake into COS 5–7 and HaCaT cells. Cells (2×10^6/mL) were exposed to LEF (20 V/cm, 500 Hz, and pulse width 180 μs) for 1 min in the presence of 1 mg/mL HRP in DMEM-H medium. Three hours after exposure, the cells were washed three times in PBS and disrupted by five freeze–thaw cycles in 50 μL PBS containing anti-protease cocktail (1:100). Dot-blot analysis of HRP activity: *1* – cells without HRP; *2* – non-exposed cells with HRP; *3* and *4* – duplicates of cells exposed to LEF in the presence of HRP.

    (2×10^6 cells/mL) are preincubated with 1 mg/ml HRP in DMEM-H for 5 min, then exposed for 1 min to LEF (20 V/cm, 180 μs, and frequency of 500 Hz). Activity of HRP in cell extracts of COS 5–7 and HaCaT cells is analyzed by Enhanced Chemiluminescence Detection Kit (e.g., SuperSignal from Pierce Biotechnology, Rockford, IL, USA) employing a dot-blot analysis 3 h after exposure. Cell extracts obtained by subjecting the cells to five freeze–thaw cycles are applied and dried on nitrocellulose membranes. The results show 6.9-fold and 3.4-fold increase in HRP activity in extracts of exposed COS 5–7 cells and HaCaT cells, respectively, relative to control cells (incubated in presence of HRP but not exposed to LEF; **Fig. 3**).

10. When analyzing fluorescence data, two effects should be taken into account: (a) control samples exhibit a rise in fluorescence when incubated with fluorescent probes at 24°C due to constitutive uptake; and (b) cells exposed to LEF in the absence of fluorescent probe, in certain culture media, reveal an increased red autofluorescence as compared to untreated control cells. This background signal needs to be subtracted from the fluorescence of all cells exposed to electric field.

Acknowledgement

This study was supported by the Israel Science Foundation founded by the Israel Academy of Sciences and Humanities (grant 1029/03 to R.K.).

References

1. Hapala, I. (1997) Breaking the barrier: methods for reversible permeabilization of cellular membranes. *Crit. Rev. Biotechnol.* **17**, 105–122.
2. Taylor, N.J. and Fauquet, C.M. (2002) Microparticle bombardment as a tool in plant science and agricultural biotechnology. *DNA Cell Biol.* **21**, 963–977.

3. Mor, G. and Eliza, M. (2001) Plasmid DNA vaccines. Immunology, tolerance, and autoimmunity. *Mol. Biotechnol.* **19**, 245–250.
4. Sundaram, J., Mellein, B.R. and Mitragotri, S. (2003) An experimental and theoretical analysis of ultrasound-induced permeabilization of cell membranes. *Biophys. J.* **84**, 3087–3101.
5. Teissie, J., Eynard, N., Gabriel, B. and Rols, M.P. (1999) Electropermeabilization of cell membranes. *Adv. Drug Delivery Rev.* **35**, 3–19.
6. Mellman, I. (1996) Endocytosis and molecular sorting. *Annu. Rev. Cell Dev. Biol.* **12**, 575–625.
7. Mukherjee, S., Richik, N.G. and Maxfield, F.R. (1997) Endocytosis. *Physiol. Rev.* **77**, 759–803.
8. Conner, S.D. and Schmid, S.L. (2003) Regulated portals of entry into the cell. *Nature* **422**, 37–44.
9. Okada, C.Y. and Rechsteiner, M. (1982) Introduction of macromolecules into cultured mammalian cells by osmotic lysis of pinocytic vesicles. *Cell* **29**, 33–41.
10. Varga, C.M., Wickham, T.J. and Lauffenburger, D.A. (2000) Receptor-mediated targeting of gene delivery vectors: insights from molecular mechanisms for improved vehicle design. *Biotechnol. Bioeng.* **70**, 593–605.
11. Qian, Z.M., Li, H., Sun H. and Ho, K. (2002) targeted drug delivery via the transferrin receptor-mediated endocytosis pathway. *Pharmacol. Rev.* **54**, 561–587.
12. Lu, Y. and Low, P.S. (2002) Folate-mediated delivery of macromolecular anticancer therapeutic agents. *Adv. Drug Delivery Rev.* **54**, 675–693.
13. Leamon, C.P. and Low, P.S. (2001) Folate-mediated targeting: from diagnostics to drug and gene delivery. *Drug Discov. Today* **6**, 44–51.
14. Rosenberg, Y. and Korenstein, R. (1997) Incorporation of macromolecules into cells and vesicles by low electric fields: induction of endocytosis-like processes. *Bioelectrochem. Bioenerg.* **42**, 275–281.
15. Antov, Y. Barbul, A. and Korenstein, R. (2004) Electroendocytosis: stimulation of adsorptive and fluid-phase uptake by pulsed low electric fields. *Exp. Cell Res.* **297**, 348–362.
16. Antov, Y. Barbul, A., Mantsur, H. and Korenstein, R (2005) Electroendocytosis: Exposure of cells to pulsed low electric fields enhances adsorption and uptake of macromolecules. *Biophys. J.* **88**, 2206–2223.
17. Entin, I., Plotnikov, A., Korenstein, R. and Keisari, Y. (2003) Tumor growth retardation, cure and induction of antitumor immunity in B16 melanoma bearing mice by low electric field enhanced chemotherapy. *Clin. Cancer Res.* **9**, 3190–3197.
18. Plotnikov, A., Fishman, D., Tischler, T., Korenstein, R. and Keisary, Y. (2004) Low electric field enhanced chemotherapy can cure mice with CT-26 colon carcinoma and induce anti tumor immunity. *Clin. Exp. Immunol.* **138**, 410–416.
19. Plotnikov, A., Tichler, T., Korenstein, R. and Keisari, Y. (2005) Involvement of the immune response in the cure of metastatic murine CT-26 colon carcinoma by low electric field enhanced chemotherapy. *Int. J. Cancer* **117**, 816–824.
20. Plotnikov, A., Niego, B., Ophir, R., Korenstein, R. and Keisari, Y. (2006) Effective treatment of mouse metastatic prostate cancer by low electric field enhanced chemotherapy. *Prostate* **66**, 1620–1630.

11

In Vitro Systems for Studying Epithelial Transport of Macromolecules

Nicole Daum, Andrea Neumeyer, Birgit Wahl, Michael Bur, and Claus-Michael Lehr

Summary

Biological barriers, typically, represented by epithelial tissues are the main hindrance against uncontrolled uptake of a variety of substances. However, the delivery across a biological barrier is a crucial factor in the development of drugs. As the permeability of macromolecular drugs is very limited, new delivery strategies have to be developed and further improved. Thereby, nanoparticle carriers offer an enormous potential for the controlled delivery of active substances into the organism. Besides an intensive study for the reason of risk assessment and toxicology, the possible transport enhancement caused by nanoparticles must be quantified. A powerful tool for these studies is in vitro cell culture models imitating the more complex in vivo situation under controlled conditions. We use polyethylenimine as model enhancer mimicking toxicological effects and altered barrier function in the epithelial in vitro model, Calu-3. Cytotoxicity assays based on different mechanisms and transport properties of a low-permeability marker with and without delivery enhancer are described.

Key Words: Biological barriers; Epithelia; Drug transport; In vitro model; Calu-3; Caco-2; Delivery enhancer; Nanoparticles; Transwell®.

1. Introduction

Human body interacts with its environment via many different ways. In this context, all surfaces that are in contact with the surrounding have to fulfil a selective function for the organism, and therefore represent a control organ or barrier. These biological barriers are the main hindrance against uncontrolled

From: *Methods in Molecular Biology, vol. 480: Macromolecular Drug Delivery*, Edited by: M. Belting
DOI 10.1007/978-1-59745-429-2_11, © Humana Press, a part of Springer Science+Business Media, LLC 2009

uptake of substances. The barrier function is mainly guaranteed by cell junctions that are especially abundant in epithelial tissues, the tight junctions *(1)*. The largest epithelia are gastrointestinal mucosa and lung with about $140 \, m^2$, while skin, cornea of the eye and others cover only smaller surfaces of several square meters or even square centimeters. As intestine and alveolar tract of the lung are only comprised of single cell layers, they are best suited to a controlled uptake of substances. While most small molecules are absorbed via passive transcellular diffusion, special carrier systems allow an active transcellular transport of others. Paracellular transport between two cells is mostly used by hydrophilic substances *(2)*. However, the presence of tight junctions avoids the permeability for larger molecules. Vesicular transport via endocytosis facilitates the entrance of rather large molecules into the epithelial cells. These transport systems are not only accessible to nutrients, but also to drug molecules; their ability to cross such a biological barrier is crucial to reach their site of action. Recent advances in computer-aided drug design, combinatorial chemistry, and genomics have generated an unequalled number of compounds needed for testing and have led to an increasing requirement for higher assay throughput. Within the development of drugs' first prerequisites are safety and efficacy, however, permeability across biological barriers and physico-chemical parameters such as solubility are important factors and cannot be neglected. While classic drug compounds (e.g., aspirin) show good permeability, the novel compounds are low permeable macromolecules belonging to the biopharmaceutics classification system (BCS) class III (low permeability, high solubility) or class IV (low permeability, low solubility) substances *(3)*. Due to their low bioavailability, these compounds demand the development and improvement of new drug delivery strategies. Such strategies include the binding and uptake to and transport through the cell or reduction of the barrier function. As nanoparticles are able to interfere with the biological barrier function of epithelia, their application as permeability enhancers for low permeable drugs is of special interest. By use of nanoparticulate carriers, an increased availability of transported drugs may take place in the lung after inhalation, through the nasal mucosa directly into the brain *(4)*, when orally administered [e.g., insulin *(5)*], and after topical application *(6–8)*.

Classic in vitro cell culture models were used earlier to test the permeability of drugs themselves *(9)*. Nowadays, these systems display not only an alternative to animal testing in respect to ethical concerns, but are also applied as advanced models to develop new delivery strategies and to study drug delivery properties of nanoparticulate carriers or chemical delivery enhancers. Thereby, interactions with the biological barrier, specifically binding, uptake, and transport can be evaluated in a well-defined and standardized system. While oral delivery of drugs is still the commonly used and most accepted method of choice due to the high compliance of patients, inhalation of medicinal aerosols to the lung for the systemic drug delivery is developing into a promising alternative to conventional routes of administration. Two-model systems, the epithelial cell lines Caco-2 *(10–13)* and Calu-3 *(14–16)* display in vitro cultures of intestinal and bronchial mucosa, respectively, and are nowadays widely accepted as standard models in the pharmaceutical industry. Their structural and biochemical characteristics together with their easy handling raise the possibility to establish these cultures even in a beginners' lab. However, more sophisticated and further developed models based on primary human cells demand higher expertise in cell handling and cultivation *(17)*. The important step for all epithelial models is the use of permeable supports in vitro. The so-called Transwell® system allows epithelial cells to be grown and studied in a polarized state under more natural conditions. Cellular differentiation can proceed to higher levels resulting in cells that morphologically and functionally closely resemble their in vivo counterparts, while cellular functions such as transport, adsorption, and secretion can be easily studied.

2. Materials
2.1. Cell Culture

1. Minimal essential medium: MEM, with Earle's balanced salts, with L-glutamine (Gibco/Invitrogen, Karlsruhe, Germany) supplemented with 10% fetal bovine serum (FBS; Sigma-Aldrich, Schnelldorf, Germany), 55 mg sodium pyruvate (Lonza, Basel, Switzerland), and 1% non-essential amino acids (NEAA; PAA, Pasching, Austria); store at 4°C.
2. Solution of trypsin (0.05%) and ethylene diamine tetraacetic acid (EDTA) (Gibco/Invitrogen, Karlsruhe, Germany).
3. Phosphate buffered saline (PBS) pH 7.4: 129 mM NaCl, 2.5 mM KCl, 7.4 mM Na_2HPO_4, 1.3 mM KH_2PO_4 (*see* **Note 1**); store at 4°C.
4. Flat bottomed, optically clear 96-well plate (Greiner, Frickenhausen, Germany).
5. Transwell® filters (Corning Costar, Bodenheim, Germany): polyester inserts for 12-well plates with a diameter of 12 mm and a pore size of 0.4 μm.
6. Polyethylenimine (PEI), 25 kDa, 10% stock solution in aqua$_{deion}$, diluted with growth medium (Sigma-Aldrich, Schnelldorf, Germany).

2.2. MTT Assay

1. MTT assay [3-(4,5-dimethylthiazol-2-yl)-2,5-diphenyltetrazolium bromide] (Roche, Mannheim, Germany).
2. PBS (*see* **Subheading 2.1**).
3. Dimethyl sulfoxide (DMSO; Fluka, Schnelldorf, Germany).
4. Tecan SLT Spectra (Tecan Deutschland GmbH, Crailsheim, Germany), 550 nm filter.

2.3. LDH Assay

1. LDH cytotoxicity detection kit (Roche, Mannheim, Germany).
2. Tecan SLT Spectra (Tecan Deutschland GmbH, Crailsheim, Germany), 492 nm filter.
3. PBS (*see* **Subheading 2.1.**).
4. Triton X-100 (Sigma-Aldrich, Schnelldorf, Germany).
5. 96-well plate (*see* **Subheading 2.1.**).

2.4. Transepithelial Electrical Resistance (TEER) Measurement

1. Volt–ohm meter (EVOM-G, World Precision Instruments, Sarasota, FL, USA) with corresponding "chopstick" electrodes.

2.5. Transport Experiments

1. Fluorescein isothiocyanate-dextrans (FITC-dextrans), 167 kDa (Synopharm GmbH & Co KG, Barsbüttel, Germany).
2. Hank's balanced salt solution (HBSS), HEPES buffer: 136.9 mM NaCl, 5.4 mM KCl, 4.26 mM NaHCO$_3$, 0.35 mM KH$_2$PO$_4$, 5.5 mM glucose, 1.26 mM CaCl$_2$, 10 mM HEPES, 0.5 mM MgCl$_2$, 0.4 mM MgSO$_4$ (*see* **Note 2**); store at 4°C.
3. CytoFluor II (Perkin–Elmer, Rodgau-Jügesheim, Germany), Software CytoFluor Version 4.2, excitation 480 nm, emission 530 nm.

3. Methods

Several cytotoxicity assays based on different mechanisms and measurement principles have been described in the literature. During safety assessment of new compounds, it is crucial to select a cytotoxicity assay addressing the correct mechanism. In this set of experiments, we use PEI *(18–20)* as model enhancer mimicking a possible effect of nanoparticles/delivery enhancers on cytotoxicity and epithelial barrier function. PEI polymers are widely used for non-viral gene delivery. For these kinds of enhancers, a cell membrane perforating mechanism is described in the literature. The rather high toxicity of PEI polymers is one of the major limiting factors especially for its in vivo use.

As recently reported, nanoparticulate carriers or other substances may interact with cytotoxicity measurement principles (*see* **Note 3**). Before performing cytotoxicity assays, it is therefore recommended to clarify whether there are any interactions between nanoparticles/delivery enhancers and test reagents. For MTT assay, it has been described that nanotubes interact with MTT-formazan crystals, therefore, the assay cannot be performed properly *(21)*. In addition, other compounds may interfere by forming a precipitate or by their color and/or intrinsic fluorescence. Taken together, the described assays can only be a suggestion depending on the substances to be tested.

All experiments that are exemplified here were performed using the human tumor cell line Calu-3, as model for the bronchial tract of the lung. The protocols are easily adaptable to other epithelial cells or cell lines taking into account that cultivation and experimental conditions (growth medium, cell numbers, cultivation time, TEER values, and others) may vary.

3.1. Cell Culture

1. Calu-3 cells are cultivated in MEM with 10% FBS, 1% non-essential amino acids (NEAA), and 55 mg sodium pyruvate. When reaching confluence, cells are washed twice with PBS and then passaged with trypsin/EDTA to obtain new maintenance cultures and experimental cultures. Fresh medium must be provided every other day (*see* **Note 4**).
2. Cells for cytotoxicity testing are seeded in a 96-well plate (10,000 cells/well) and grown for 4 days in a final volume of 100 µL cell culture medium/well in a humidified atmosphere with 5% CO_2 and a temperature of 37°C.
3. Transwell cultures are prepared by adding 500 µL of cell suspension (concentration of 200,000 cells/mL, i.e., 100,000 cells/well) to the apical compartment of a Transwell® system, and by adding 1.5 mL of medium to the basolateral compartment. Transwells® are then incubated for 9–12 days at 37°C and 5% CO_2 in the incubator until a confluent cell monolayer is formed on the filter membrane. The cellular resistance across the monolayer steadily increases as the cells differentiate. This time, TEER values are documented under sterile conditions (*see* **Subheading 3.4.**). Cultures with TEER values of 600–1200 Ω-cm^2 are used for the experiments.

3.2. MTT Assay

MTT assay is a standard colorimetric assay used to determine cytotoxicity of potential medicinal agents and other toxic materials. It is based on the reduction of the tetrazolium salt MTT by viable cells. A mitochondrial dehydrogenase enzyme is able to cleave the tetrazolium rings of the pale yellow MTT and form dark purple formazan crystals, which are largely impermeable to cell membranes resulting in the accumulation of these crystals within healthy cells. Solubilization of the cells by the addition of a detergent results in the liberation of crystals, which are solubilized. The metabolic activity of cells is directly proportional to the concentration of the created formazan product (*22*), whose color is quantified in a colorimetric assay.

1. Incubate the cells with the designated test solution (e.g., nanoparticles/delivery enhancers, in our case PEI). Include also a negative control (normal culture medium, 100% metabolic activity) in triplicates to determine the maximum value.

The incubation period depends on the particular experimental approach; PEI and control are incubated for 4 h.

2. After the incubation period, 10 μL of the MTT solution (5 mg MTT per mL PBS) is added to each well (*see* **Notes 5 and 6**).
3. The microplate has to be incubated for another 4 h in a humidified atmosphere (e.g., 37°C, 5% CO_2), then the supernatant is removed.
4. Fifty microliters of DMSO is added to each well to solubilize the crystals.
5. Incubate the plate for about 20 min at room temperature protected from light. Check for complete solubilization of the purple formazan dye.
6. Mix the plate for about 1 min on a shaker.
7. Measure the spectrophotometrical absorbance of the samples using a microplate (ELISA) reader at a wavelength of 550 nm. The reference wavelength should be more than 650 nm.
8. Concentration of PEI is plotted against metabolic activity, and sigmoidal fitting is performed to determine the EC_{50} value (*see* **Fig. 1**).

3.3. LDH Assay

Membrane damage of cells can be used as a principle to determine the cytotoxicity of compounds or nanoparticles. The terminal stages of cell death lead to damage of cell membranes and to an irreparable loss of cell membrane integrity. This results in the leakage of certain cytoplasmic enzymes, such as adenylate

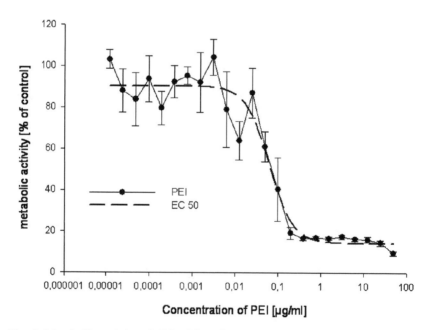

Fig. 1. Metabolic activity of Calu-3 is reduced by PEI, EC_{50} value $= 0.062\ \mu g/mL$, $n = 6$.

kinase or lactate dehydrogenase (LDH), into the cell culture supernatant. By the determination of the catalytic activity of LDH in the supernatant of a cell culture, it is therefore possible to determine the cytotoxic effects of nanoparticles or delivery enhancers.

1. Buffer (PBS or HBSS, alternatively) should be prewarmed to 37°C.
2. Prepare a dilution series of the compounds you want to test for cytotoxic effects in prewarmed buffer, store them in a 37°C water bath until use. For each compound concentration to be tested at least 650 μL is necessary (i.e., 3 × 200 μL for a triplicate approach).
3. To determine the maximal amount of LDH present in the supernatant, use 1% Triton-X in buffer as a positive control to kill all cells (*see* **Note 7**).
4. Remove the culture medium; be careful not to disturb the cell layer. Wash cells twice with 200 μL of prewarmed HBSS and suck off the buffer carefully.
5. Add 200 μL of each compound concentration into triplicate wells. Add 200 μL of 1% Triton-X in buffer as positive control into each of the triplicate wells as well as 200 μL of buffer as negative control, and incubate the plate for 4 h on a shaker (rotation frequency of ~200 rpm) in an incubator (37°C, 5% CO_2, 90% humidity) (*see* **Note 8**).
6. Prewarm the bottle 2 from the assay kit in a 37°C water bath (*see* **Note 9**).
7. A few minutes before the end of the incubation time prepare the working solutions: The kit contains two working solutions. Solution 1 contains the catalyst. When using the assay kit for the first time, it is necessary to reconstitute the lyophilizate for 10 min by adding 1 mL of aqua$_{deion}$ into bottle 1 and mixing it thoroughly. The reconstituted catalyst may be stored at 4°C for several weeks. Bottle 2 contains the dye solution and is ready to use. To prepare reaction solution for 100 tests, mix 11.25 mL of dye solution (bottle 2) with 250 μL of catalyst (bottle 1). This reaction mixture should not be stored, so prepare it immediately before use (*see* **Note 10**).
8. Transfer 100 μL of supernatant from each well to the probe sampling plate. Be careful not to damage the cells at the bottom of the well.
9. Add 100 μL of reaction mixture to each well and incubate for ~5 min at room temperature. During the incubation time keep the 96-well plate in the dark. A color reaction takes place and the dark red formazan dye will be formed (*see* **Note 11**).
10. Measure absorbance at 492 nm with a reference wavelength of more than 600 nm.
11. Normal culture medium is used as negative control (0% LDH); Triton-X is used as positive control (100% LDH). Concentration of PEI is plotted against LDH release, and sigmoidal fitting is performed to determine the EC_{50} value (*see* **Fig. 2**).

3.4. TEER Measurement

The quality of barrier properties can be measured by the TEER value. Several compounds are able to increase paracellular transport only by opening the tight junctions, thus influencing the cell layer integrity. Other substances, such as

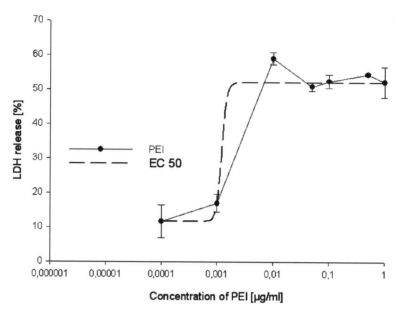

Fig. 2. LDH release of Calu-3 is increased by PEI, EC_{50} value $= 0.0012\,\mu g/mL$, $n = 8$.

PEI, reduce the barrier properties by perforating the cell membrane and thereby affecting the integrity of the cells themselves. The amount of resistance degradation of both mechanisms can be determined by TEER.

1. The resistance is measured by use of an epithelial volt–ohm meter (EVOM) with corresponding "chopstick" electrodes. It is very important to handle the electrodes with care, and to never touch the cell monolayer.
2. To calibrate the EVOM, it is necessary to determine the resistance of the HBSS buffer at the same temperature as used in the experiment. Dip the chopstick electrodes into the buffer, and press the "test" button to determine the TEER value of the blank buffer. If the obtained value differs from 0, calibrate the EVOM by turning the calibration screw with a screwdriver till the value 0 is reached for the blank buffer (*see* **Note 12**).
3. After calibration, the long side of the electrode is dipped into the basolateral compartment, the short into the apical compartment. The TEER value of every well has to be noted. The measured TEER values must be normalized to the filter area ($1.13\,cm^2$). Correction for the surface area is done by multiplying the measured value with 1.13.
4. Cells are incubated with different concentrations of PEI for 4 h. Normal culture medium is used as negative control (100% TEER value). Concentration of PEI is plotted against the change of TEER, and sigmoidal fitting is performed to determine the EC_{50} value (*see* **Fig. 3**).

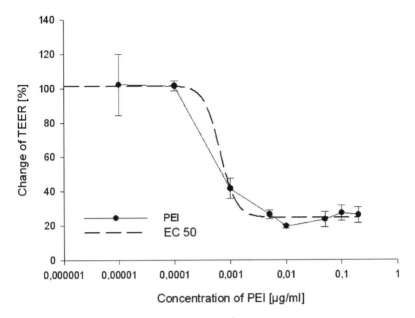

Fig. 3. TEER values of Calu-3 on Transwell® filters (pore size 0.4 μm) are reduced by PEI, EC_{50} value = 0.0007 μg/mL, $n = 6$.

3.5. Transport of a Low Permeability Marker

New drug delivery strategies include the binding, uptake, and transport through the cell by special carrier systems or reduction of the barrier function. Comparing the transport of a low permeability marker such as FITC-dextrans with and without delivery enhancer indicates to what extent this transport can be influenced.

1. HBSS and test solutions are prewarmed in a 37°C water bath.
2. Before starting the transport experiment, the TEER values have to be measured to control the integrity of the cell monolayer (compare **Subheading 2.4.**, *see* **Note 13**).
3. Cell monolayers grown on Transwell® filters have to be washed twice with freshly prepared and prewarmed HBSS. To avoid damage of the monolayer by hydrostatic pressure, first the medium from the basolateral compartment and then from the apical compartment has to be removed. The apical compartment is filled with 500 μL of fresh buffer and then the basolateral compartment with a volume of 1500 μL.
4. Incubate the Transwell® chamber for 60 min at 37°C and 5% CO_2 with HBSS in the incubator (*see* **Note 14**).
5. After preincubation with HBSS, TEER values have to be measured again. Due to the change of the medium, which is associated with the removal of important

nutrients, the TEER values will be slightly reduced compared to the values of the first measurement.

6. Remove the buffer and add 520 µL of the test solution (FITC-dextrans with 0.01% PEI and without PEI) to the upper (apical) compartment of the first well and 1500 µL HBSS buffer to the lower (basolateral) compartment of the first well to measure the transport from apical to basolateral (AB).

7. Directly after the addition of the test solution, take a sample of 20 µL from the donor compartment to determine the exact starting concentration and pipette it into a volume of 180 µL HBSS buffer, which has been provided on a 96-well plate.

8. Incubate the Transwell® chamber at 37°C and 5% CO_2 in the incubator on a shaker with a constant rotation speed.

9. Take aliquots of 200 µL from the acceptor compartment at designated time points (e.g., 30, 60, 90, 120, 150, and 180 min). Transfer the volume to a 96-well plate and replace the removed volume by pipetting 200 µL of fresh HBSS buffer into the acceptor compartment.

10. After the last sampling, take a sample of 20 µL from the donor compartment again to determine the final concentration and transfer it to a volume of 180 µL HBSS buffer, which has been provided in a 96-well plate.

11. Analyze the samples from both donor and acceptor compartments by fluorescence measurement. Measure FITC-dextrans via CytoFluor plate reader, set excitation at 480 nm and emission at 530 nm. It is not necessary to perform a measurement after each sampling. Hence, it is possible to store the 96-well plate with the samples at room temperature protected from light. Perform one measurement of the plate for all time points together at the end of the experiment.

12. Plot the standard curve (concentration versus measured fluorescence values and peak areas) based on the standard concentration of the tested substances. Perform linear regression.

13. Use the parameters of the standard curve to calculate the concentrations of the different samples. Note that the fluorescence values and peak areas, of the donor samples (determination of the starting and end concentrations), must be multiplied by ten due to the dilution step when taking the sample.

14. Calculate the cumulative amount (in mmol) of the compound transported from the donor to the acceptor compartment at each time point. The amounts that are taken away from the acceptor compartment at each sampling time must be added back.

15. Plot the cumulative amount versus time and use the linear portion of the plot to determine the appearance rate ($\Delta Q/\Delta t$, in mmol/s) according to the following equation:

$$P_{app} = (\Delta Q/\Delta t)(60 \times A \times C_0)$$

A is the surface area of the cell monolayer in cm^2, c_0 is the initial concentration of the compound on the donor side in $mmol/cm^3$. When the transported amount

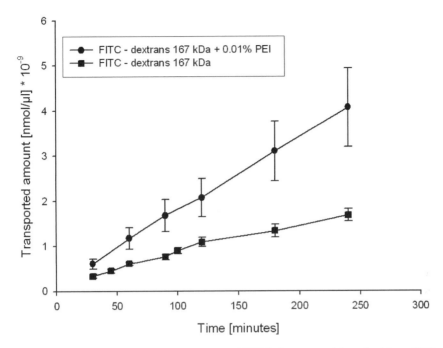

Fig. 4. Comparison of the transport rates of FITC-dextrans with and without PEI as delivery enhancer across a Calu-3 monolayer on Transwell® filters (pore size 0.4 μm). Transport was documented over a period of 250 min.

of the substrate is plotted against the time, a regression line can be applied for the linear part of the curve. The slope of this regression line corresponds to the flux across the monolayer, the steady-state transport under sink conditions (donor concentration at the end of the experiment >90% of c_0).

16. The p_{app} values of the two setups mirror their different transport properties. While FITC-dextrans alone are transported in a rate of 2.78×10^{-8} cm/s; FITC-dextrans in combination with the delivery enhancer PEI are transported significantly faster with 1.32×10^{-7} cm/s (*see* **Fig. 4**).

4. Notes

1. A premix of the dry ingredients of PBS can be prepared. The premix should consist of 7.54 g/L NaCl, 0.186 g/L KCl, 1.896 g/L Na_2HPO_4, and 0.177 g/L KH_2PO_4. It is possible to prepare the premix for several liters of ready to use PBS and keep the dry ingredients well sealed at room temperature. To prepare 1 L of PBS, dissolve 9.799 g of the premix in aqua$_{deion}$ and adjust pH to 7.4.

2. To minimize errors occurring when weighing the compounds, HEPES-buffered HBSS may be prepared from three stock solutions. Prepare 1 L 10× stock no. 1

with 1.369 M NaCl, 54 mM KCl, 42.6 mM NaHCO$_3$, 3.5 mM KH$_2$PO$_4$, 55.5 mM glucose, and 100 mM HEPES, 100 mL 100× stock no. 2 with 126 mM CaCl$_2$, and 100 mL 100× stock no. 3 with 50 mM MgCl$_2$ and 41 mM MgSO$_4$. To prepare 1 L working solution dilute 100 mL stock no. 1 in 850 mL aqua$_{deion}$, add 10 mL stock no. 2 and 10 mL stock no. 3, adjust pH to 7.4, and fill up to 1 L with aqua$_{deion}$. The stock solutions can be stored at 4°C.

3. As literature has reported recently, certain nanoparticles/nanotubes may interact with this cytotoxicity measurement principle. To exclude the possibility of compound–assay interactions, it is highly recommended to perform substance controls. Such controls can be conducted by performing the assay according to the given protocol, but without cells. This will exclude that the particles give a wrong "positive signal" in the assay.

4. To get a weekly rhythm, cells may also be fed in a 2:2:3 days rhythm.

5. It is recommended to dissolve a larger amount of MTT in PBS and make aliquots of it. MTT solution can be stored at –20°C for several months. After thawing, the reagent may be stored protected from light at +2 to +8°C for up to 4 weeks.

6. If a larger volume of culture medium is required, the amount of MTT solution has to be increased correspondingly.

7. Due to high viscosity, Triton-X is very difficult to pipette. So, it is recommended to cut off a part of the pipette tip to ensure an easier pipetting.

8. By using a 96-well plate with cells, it is also possible to determine the cytotoxic effects in a time-dependent manner by using a larger amount of wells per tested concentration and by harvesting the supernatant at different time points. So, it is possible to receive time-dependent as well as concentration-dependent statistics.

9. As the kit is stored at –20°C and bottle 2 of the kit contains a large volume of dye solution, you should take care that you have enough time to thaw and prewarm the solution in a water bath. Calculate at least 30 min to ensure complete thawing.

10. Until further use, the dye solution may be refrozen and the rest of the reconstituted catalyst can be stored at 4°C for several weeks.

11. According to the manufacturer's instructions, an incubation time of up to 30 min is possible. In practical approaches, an incubation time of ~5 min has proven to be advantageous as the formation of the formazan salt takes place very fast resulting in a dark red color of the solution. So, absorbance easily can reach values larger than 4.0, which is the maximum absorbance the used plate reader is able to measure.

12. Instead of using a separate vial with buffer for the calibration of the EVOM, it is also possible to pipette some buffer into the interjacent slots between the wells of a Transwell® plate and measure the buffer resistance there. This guarantees also comparable conditions (temperature, etc.).

13. Cell cultivation has to be performed under sterile conditions. However, the transport experiment itself can be performed under non-sterile conditions due to its short duration of several hours.

14. Also, other buffers with similar contents can be used. The important step is the addition of calcium and magnesium ions, due to the fact that the tight junctions

are (Ca^{2+})- and (Mg^{2+})-dependent. A buffer without these ions will lead to opening of the tight junctions resulting in a significantly reduced TEER value and higher paracellular permeability.

References

1. Cerijido, M. (ed.) (1992) *Tight Junctions*. CRC Press, Boca Raton.
2. Hayashi, M. and Tomita, M. (2007) Mechanistic analysis for drug permeation through intestinal membrane. *Drug Metab Pharmacokinet* **22,** 67–77.
3. Amidon, G.L., et al. (1995) A theoretical basis for a biopharmaceutic drug classification: the correlation of in vitro drug product dissolution and in vivo bioavailability. *Pharm Res* **12,** 413–20.
4. Oberdorster, G., et al. (2004) Translocation of inhaled ultrafine particles to the brain. *Inhal Toxicol* **16,** 437–45.
5. Damge, C., et al. (1988) New approach for oral administration of insulin with polyalkylcyanoacrylate nanocapsules as drug carrier. *Diabetes* **37,** 246–51.
6. Luengo, J., et al. (2006) Influence of nanoencapsulation on human skin transport of flufenamic acid. *Skin Pharmacol Physiol* **19,** 190–7.
7. Alvarez-Roman, R., et al. (2004) Enhancement of topical delivery from biodegradable nanoparticles. *Pharm Res* **21,** 1818–25.
8. Toll, R., et al. (2004) Penetration profile of microspheres in follicular targeting of terminal hair follicles. *J Invest Dermatol* **123,** 168–76.
9. Wilson, G. (1990) Cell culture techniques for the study of drug transport. *Eur J Drug Metab Pharmacokinet* **15,** 159–63.
10. Artursson, P., Palm, K., and Luthman, K. (1997) Intestinal drug absorption and metabolism in cell cultures: Caco-2 and beyond. *Pharm Res* **14,** 1655–8.
11. Artursson, P., Palm, K., and Luthman, K. (2001) Caco-2 monolayers in experimental and theoretical predictions of drug transport. *Adv Drug Deliv Rev* **46,** 27–43.
12. Fogh, J., Fogh, J.M., and Orfeo, T. (1977) One hundred and twenty-seven cultured human tumor cell lines producing tumors in nude mice. *J Natl Cancer Inst.* **59,** 221–6.
13. Hilgers, A.R., Conradi, R.A., and Burton, P.S. (1990) Caco-2 cell monolayers as a model for drug transport across the intestinal mucosa. *Pharm Res.* **7,** 902–10.
14. Florea, B.I., et al. (2003) Drug transport and metabolism characteristics of the human airway epithelial cell line Calu-3. *J Control Release* **87,** 131–8.
15. Lehr, C.M., Bur, M., and Schaefer, U.F. (2006) Cell culture models of the air-blood barrier for the evaluation of aerosol medicines. *ALTEX* **23 Suppl,** 259–64.
16. Grainger, C.I., et al. (2006) Culture of Calu-3 cells at the air interface provides a representative model of the airway epithelial barrier. *Pharm Res* **23,** 1482–90.
17. Ehrhardt, C., Kim, K.J., and Lehr, C.M. (2005) Isolation and culture of human alveolar epithelial cells. *Methods Mol Med* **107,** 207–16.
18. Campeau, P., et al. (2001) Transfection of large plasmids in primary human myoblasts. *Gene Ther* **8,** 1387–94.
19. Fischer, D., et al. (1999) A novel non-viral vector for DNA delivery based on low molecular weight, branched polyethylenimine: effect of molecular weight on transfection efficiency and cytotoxicity. *Pharm Res* **16,** 1273–9.

20. Marschall, P., Malik, N., and Larin, Z. (1999) Transfer of YACs up to 2.3 Mb intact into human cells with polyethylenimine. *Gene Ther* **6,** 1634–7.
21. Worle-Knirsch, J.M., Pulskamp, K., and Krug, H.F. (2006) Oops they did it again! Carbon nanotubes hoax scientists in viability assays. *Nano Lett* **6,** 1261–8.
22. Mosmann, T. (1983) Rapid colorimetric assay for cellular growth and survival: application to proliferation and cytotoxicity assays. *J Immunol Methods* **65,** 55–63.

12

Preparation of Macromolecule-Containing Dry Powders for Pulmonary Delivery

Kelly S. Kraft and Marshall Grant

Summary

Drug delivery by inhalation is routine for the treatment of local pulmonary conditions like asthma, cystic fibrosis, and chronic obstructive pulmonary disease. Only recently, though, has the inhalation route been considered for administering drugs for systemic diseases. The pulmonary route is attractive for several reasons. It is non-invasive, it avoids first-pass metabolism, and it allows drug absorption from a large, highly vascularized surface area. However, consistent delivery to the deep lung requires drug particles within a very narrow size range. Several particle engineering approaches have been used to produce dry powders that will reach the alveolar space. Some of these methods, such as spray drying from solution, the formation of drug-containing liposomes, and the controlled crystallization of particles, are described here.

Key Words: Dry powder inhalers (DPI); Pulmonary drug delivery; Insulin; Particle engineering; Spray drying; Liposomes; Aerosol solvent extraction system (ASES); Technosphere® insulin.

1. Introduction

Over the past two decades, advances in peptide and protein manufacturing technologies have led to the production of many novel therapeutic agents *(1)*. These new drug candidates are commonly administered by injection to avoid the complexities associated with either oral or pulmonary delivery. Proteolysis in the digestive tract is the primary obstacle to oral protein delivery, and reproducible delivery to the deep lung is the primary obstacle to pulmonary peptide

From: *Methods in Molecular Biology, vol. 480: Macromolecular Drug Delivery,* Edited by: M. Belting
DOI 10.1007/978-1-59745-429-2_12, © Humana Press, a part of Springer Science+Business Media, LLC 2009

administration. Despite these significant challenges, however, and to eliminate frequent injections and increase patient compliance, formulations that allow the non-invasive administration of peptide and protein therapeutics are an active area in drug development.

While drug delivery by inhalation is routine for the treatment of local pulmonary conditions like asthma, cystic fibrosis, and chronic obstructive pulmonary disease *(2,3)*, only recently has the inhalation route been considered to administer drugs for systemic diseases. The pulmonary route is attractive because the deep lung provides a very large, highly vascularized surface area for drug absorption. In addition, protein and peptide degradation is minimized due to the relatively limited number of proteolytic enzymes found in the lung fluid, and drugs absorbed through the lung avoid first-pass metabolism. The first inhaled drug approved for the treatment of a non-pulmonary disease is EXUBERA®, an inhaled insulin for adults with type 1 or 2 diabetes mellitus.

The primary challenge in pulmonary drug administration is reproducible delivery of therapeutic agents to the deep lung. The deep lung is targeted for drug absorption because the alveoli represent the site of maximum surface area and, thus, maximum drug absorption. Consistent delivery to the alveoli requires the preparation of drug particles within a very narrow size range. Particles for pulmonary delivery are classified by aerodynamic diameter, d_a, which is related to the terminal sedimentation velocity in air of an equivalent sphere with a density (ρ_a) of 1 g/cm^3. The relationship between a particle's physical diameter, d_p, and its aerodynamic diameter is $d_a = d_p\sqrt{\rho_p/\rho_a}$, where ρ_p is the effective particle density. Thus, the particle acts aerodynamically larger than its physical size when $\rho_p > \rho_a$, and aerodynamically smaller when $\rho_p < \rho_a$.

Particles appropriately sized to reach the deep lung upon inhalation are defined as being in the respirable range or respirable fraction (0.5–5.8 μm). In general, the inertia associated with particles having an aerodynamic diameter larger than 10 μm prevents them from following the pulmonary streamlines during inhalation. As a result, these particles deposit in the throat and are swallowed. Particles between 6 and 10 μm deposit in the conducting airways (bronchi and bronchioles) and cannot be absorbed through the thick epithelial layers of these airways. Particles between 0.5 and 5.8 μm are appropriately sized to follow the inhalation streamlines into the alveolar regions. It is this size range that is optimal for inhaled therapeutic agents. Particles smaller than 0.5 μm have insufficient inertia to reach the lung wall and are typically exhaled (**Fig. 1** and Color Plate 3, *see* Color Plate Section) *(3)*.

Three different types of inhalers are used to deliver drugs to the deep lung. Nebulizers, the oldest type of inhaler, work by producing a mist of aqueous, drug-containing droplets. The drug can be either dissolved or suspended in the water phase. In a pressurized metered dose inhaler (pMDI), the drug is

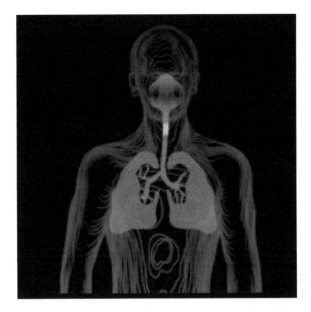

Fig. 1. Particle deposition in the lungs. Particles larger than 10 μm are deposited in the mouth and throat (*yellow/orange area*) and are swallowed. Particles between 6 and 10 μm are deposited in the upper airways (*blue area*). Particles between 0.5 and 6 μm are within the respirable range and are deposited in the alveolar region (*pink area*). Image from:http://www.filterair.info/articles/article.cfm/ArticleID/36856F0C-747B-4E08-B730798D614269E9/Page/1 (*see* Color Plate 3, *see* Color Plate Section).

either dissolved or suspended in a liquid propellant (hydrofluoroalkane). The propellant quickly evaporates after it leaves the pMDI, depositing drug micropar-ticles in the lung.

Dry powder inhalers (DPIs) administer the drug to the patient as a dry powder. DPIs offer the advantage of improved stability of the drug powder relative to a solution or a liquid suspension. In addition, DPIs can be used to administer higher drug doses than nebulizers and pMDIs. DPIs, however, also impose con-straints on the drug product. The manufacturing conditions must reproducibly yield powder particles in the 0.5–5.8 μm size range. An added complication resides in the tendency of these small particles to aggregate into particles that are too large to reach the deep lung. Thus, the inhaler device must also be engineered to deagglomerate the aggregated powders. Consequently, DPI devices tend to be product-specific.

Several approaches have been used to produce dry powders that will reach the alveolar space. The most direct methods of manufacture involve produc-ing geometrically and aerodynamically small particles by spray drying from solution, through the formation of drug-containing liposomes, or by controlled crystallization of particles composed of active pharmaceutical ingredients and carrier molecules. Another approach takes advantage of the effect of particle density on aerodynamics by producing particles that are relatively large geometrically but behave smaller aerodynamically. These particles are, typically, produced by emulsification and supercritical fluid techniques. Examples of these approaches are described here. Where possible, the preparation of insulin-containing powders is described.

2. Materials

2.1. Macromolecules

1. Bovine insulin powder (Sigma–Aldrich, St. Louis, MO, USA); store at –20°C.
2. Crystalline egg white lysozyme (Fluka, Milwaukee, WI, USA); store at 2–8°C.
3. Albumin; bovine fraction V (Fluka, Milwaukee, WI, USA); store at 2–8°C.
4. Recombinant human deoxyribonuclease (Genentech Inc, South San Francisco, CA, USA).

2.2. Excipients

1. α-Lactose monohydrate (Acros Organics, Morris Plains, NJ, USA); store powder at ambient temperature.
2. D-Trehalose dihydrate (Sigma, St. Louis, MO, USA); store powder at ambient temperature.
3. D-Mannitol (Sigma, St. Louis, MO, USA); store powder at ambient temperature.
4. Dipalmitoylphosphatidylcholine (DPPC; Lipoid, Ludwigshafen, Germany); store at –20°C.
5. Poly(lactic acid-*co*-glycolic acid) (PLGA; Birmingham Polymers, Inc.); protect from moisture.
6. Cholesterol (Sigma, St. Louis, MO, USA); store at –20°C.
7. L-α-Phosphatidylcholine (Sigma, St. Louis, MO, USA); store at –20°C.
8. Technosphere® particles (MannKind Corporation, Danbury, CT, USA); store at ambient temperature.

2.3. Solvents, Buffers, and Reagents

1. Phosphate buffered saline (PBS), pH 7.4 (Sigma, St. Louis, MO, USA). Reconstitute powder in water as directed and store at ambient temperature.
2. Sodium citrate buffer, pH 3–5 (Fluka, Milwaukee, WI, USA); store at ambient temperature and use by expiration date on the container.
3. Methylene chloride (Sigma–Aldrich, St. Louis, MO, USA); store at ambient temperature.
4. Carbon dioxide (TechAir, Danbury, CT, USA).

5. Chloroform, containing ethanol as a stabilizer (Sigma–Aldrich, St. Louis, MO, USA); store at ambient temperature.
6. Ethanol, 200 proof (Sigma–Aldrich, St. Louis, MO, USA); store at ambient temperature.
7. Dimethyl sulfoxide (DMSO; Sigma–Aldrich, St. Louis, MO, USA); store at ambient temperature.

3. Methods

3.1. Spray Drying (4,5)

A dilute solution of insulin in an appropriate aqueous medium, such as PBS (pH 7.4) or citrate buffer (pH 3 or 5), is prepared (*see* **Note 1**). Additives, such as lactose, trehalose, mannitol, or DPPC, are added as needed. These additives serve many purposes: as bulking agents *(4)*; to improve processing *(4)*, to improve stability *(6)*, and to increase bioavailability of the administered drug *(4,7)*.

The resulting solution is then spray dried. The following parameters are all controlled during the spray-drying process: solution feed flow rate, atomization nozzle gas flow rate, inlet air temperature, and drying gas flow rate. These controlled factors all affect the mass and energy balances during spray drying and determine the outlet temperature. This outlet temperature can affect the stability of the drug product because, in most spray-drying systems (co-current flow), the dried particles are exposed to a constant flow of outlet gas. For insulin powders, this outlet air temperature should not exceed 120°C *(5)*. The particle morphology as determined by scanning electron microscopy is not sensitive to the choice of drying parameters *(5)*.

The physical size of the resulting spray-dried particles is dictated by the mass of solute in each droplet and depends on both the size of the droplets atomized into the hot air stream and the concentration of the feed solution. More dilute solutions permit the formation of larger droplets, while more concentrated solutions require smaller droplets. The appropriate feed concentrations and atomization conditions must also be practical to be commercially viable. For example, doubling the feed concentration halves the amount of water, energy, and time required to produce a unit of drug powder. Typically, the spray-drying parameters can be adjusted to produce particles within the size range needed for alveolar deposition.

3.2. Large Porous Particles

3.2.1. Emulsification Techniques (8)

An aqueous solution of insulin is added to a solution of PLGA in methylene chloride. Additives, such as L-α-phosphatidylcholine dipalmitoyl, are included as needed to increase the porosity of the powder product *(8)*. The mixture is emulsified by sonication on ice and then added to an aqueous solution of polyvinyl alcohol and homogenized. The resulting emulsion is stirred to remove the methylene chloride, and the resulting microspheres are allowed to harden. The microspheres are collected by centrifugation, washed, and dried by lyophilization. The resulting porous particles possess a mean geometric diameter of approximately 8.5 μm, but an aerodynamic diameter of 2.5 μm. Thus, approximately 50% of the powder is in the respirable range, as measured by cascade impactor.

3.2.2. Supercritical Carbon Dioxide

Supercritical carbon dioxide (CO_2) has been used to generate large porous particles *(9,10)*. In one process, small, non-porous microspherical powders generated by emulsification are sealed in a high-pressure vessel and allowed to equilibrate with CO_2 at 33°C and 1200psi (supercritical conditions for CO_2) for 30 min. This incubation period allows the powders to absorb the CO_2 and swell. The pressure is then released, and the absorbed CO_2 is released from the powder. The recovered particles possess a mean geometric diameter of 13.8 μm, six times larger than the geometric diameter of the starting powders (2.2 μm), and an aerodynamic diameter of 3.7 μm.

3.3. Liposomes (11)

In conventional film-shaking liposome preparation, lipids (*see* **Note 2**) are dissolved in chloroform and the solvent removed by evaporation to leave a lipid film. Insulin-containing citrate buffer (pH 4.0) is added to hydrate the lipid film. Mechanical shaking of the mixture results in the formation of multilamellar vesicles.

In the membrane destabilizing/detergent dialyzing method, "empty" liposomes are formed by slowly adding a lipid/ethanol mixture to citrate buffer. These "empty" liposomes are extruded through two 200 nm filters to give large unilamellar liposomes. Insulin is dissolved in citrate buffer solution then added with vortexing. The resulting mixture is incubated at 40°C for 1 h to allow the insulin to be incorporated into the liposomes, and then dialyzed to remove the ethanol.

The recovered liposomes possess a mean geometric diameter of 0.2 μm. The aerodynamic diameter of these particles, when emitted from an ultrasonic nebulizer, is 1 μm.

3.4. Precipitation Techniques

3.4.1. Supercritical Carbon Dioxide

In the aerosol solvent extraction system (ASES), the protein is dissolved in a compatible solvent (i.e., water or DMSO) and then introduced by atomization into supercritical CO_2 *(12,13)*. The solvent is extracted from the droplet, and the protein precipitates to form particles with appropriate sizes for pulmonary delivery (*see* **Note 3**). CO_2 is then removed by venting off and the particles are collected. Insulin powders (*see* **Note 4**) made by this method possess a mean geometric diameter of 9.6 μm, and 22% of the powder was in the respirable range (0.5–6 μm) *(12)*.

3.4.2. Technosphere® Insulin Technology

Technosphere® Insulin (TI) inhalation powder is prepared from fumaryl diketopiperazine (FDKP), a diketopiperazine derived from lysine and functionalized with fumaric acid groups on each side arm of the molecule (**Fig. 2**).

FDKP is insoluble in acid, but is soluble under neutral conditions where the acid protons are readily dissociated. FDKP is dissolved in dilute aqueous base and then crystallized by acidification. The resulting FDKP crystals self-assemble into structures with an overall spherical morphology (**Fig. 3**), high-specific surface area, and open architecture. The size and surface area of these Technosphere® particles can be adjusted by controlling the crystallization conditions.

A solution of insulin is added to an aqueous suspension of Technosphere® particles. The insulin is adsorbed to the surface of the particles to form TI. The solvent is then removed from the suspension by lyophilization to produce the bulk drug powder. TI has a mean geometric diameter of 2.0 μm and an aerodynamic diameter of 2.5 μm.

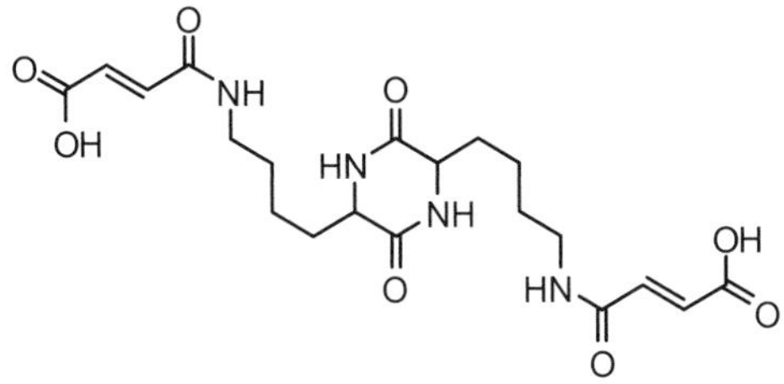

Fig. 2. Chemical structure of FDKP.

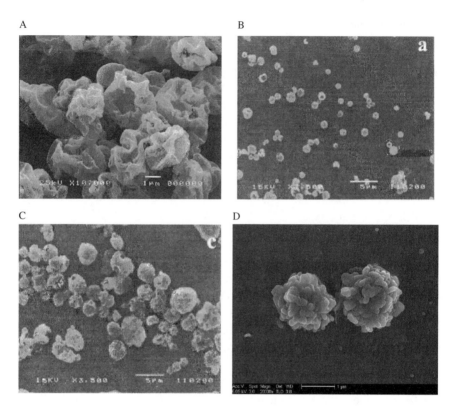

Fig. 3. SEM images of various dry powders for inhalation. (**a**) Spray-dried particles (*see* **Subheading 3.1.**). Reproduced from **ref. 5**. (**b**) Particles prepared by emulsification techniques (*see* **Subheading 3.2.1.**). Reproduced from **ref. 10**. (**c**) Particles prepared by supercritical CO_2 swelling (*see* **Subheading 3.2.2.**). The particles in panel b were the starting material for these particles. Reproduced from **ref. 10**. (**d**) TI particles (*see* **Subheading 3.4.2.**).

3.5. Aerodynamic Characterization

Aerodynamic particle size distributions are measured with instruments known as cascade impactors that classify particles into specified size "bins." Several types of impactors are available, for use over appropriate ranges of flow rates. All are based on the operating principle of inertial impaction: the aerosol stream passes through a series of "stages" in which the air stream accelerates as it moves to the next stage *(14)*. Particles that are aerodynamically small enough to travel with the air stream move to the next stage, while those that are too large are captured on the current stage. The stages are calibrated in terms of an "effective cutoff diameter" that determines the size range of particles captured.

Cascade impaction can be used to measure the dose of a powder that is emitted from an inhaler (emitted dose). The drug is discharged into the impactor and the particles are collected from the various stages. The quantity of drug on each stage can be measured either gravimetrically or by an appropriate sensitive and specific assay like high-performance liquid chromatography. In cases, where it is necessary to determine the amount of drug that would be delivered in each size range, the powder collected at each stage can be assayed individually to determine the "fine particle dose" or quantity of drug contained in the respirable fraction.

4. Notes

1. For macromolecules that are water insoluble, ethanol/water mixtures can be used for solubilization.
2. Some commonly used lipids include DPPC, cholesterol, and phosphatidylcholine *(11,15)*.
3. Processing parameters, such as operating temperature, protein concentration, and the presence of antisolvents in the supercritical CO_2, influenced the extent of aggregation observed in the resulting powders.
4. **Ref.** *12* describes the preparation of microspheres of lysozyme, insulin, albumin, and recombinant human deoxyribonuclease. Of these proteins, insulin powders demonstrated the highest degree of agglomeration.

References

1. Agu, R.U., et al. (2001) The lung as a route for systemic delivery of therapeutic proteins and peptides. *Respiratory Research*, **2**: p. 198–209.
2. Laube, B. (2005) The expanding role of aerosols in systemic drug delivery, gene therapy, and vaccination. *Respiratory Care*, **50**: p. 1161–1176.
3. Clark, A.R., et al. (2006) The application of pulmonary inhalation technology to drug discovery. *Annual Reports in Medicinal Chemistry*, **41**: p. 384–393.
4. Todo, H., et al. (2001) Effect of additives on insulin absorption from intratracheally administered dry powders in rats. *International Journal of Pharmaceutics*, **220**: p. 101–110.
5. Stahl, K., et al. (2002) The effect of process variables on the degradation and physical properties of spray dried insulin intended for inhalation. *International Journal of Pharmaceutics*, **233**: p. 227–237.
6. Bosquillon, C., et al. (2004) Aerosolization properties, surface composition and physical state of spray-dried protein powders. *Journal of Controlled Release*, **99**: p. 357–367.
7. Codrons, V., et al. (2003) Systemic delivery of parathyroid hormone (1-34) using inhalation dry powders in rats. *Journal of Pharmaceutical Sciences*, **92**: p. 938–950.
8. Edwards, D.A., et al. (1997) Large porous particles for pulmonary drug delivery. *Science*, **279**: p. 1868–1871.

9. Koushik, K., et al. (2004) Pulmonary delivery of deslorelin: large-porous PLGA particles and HPBCD complexes. *Pharmaceutical Research*, **21**: p. 1119–1126.
10. Koushik, K. and U.B. Kompella (2004) Preparation of large porous deslorelin-PLGA microparticles with reduced residual solvent and cellular uptake using a supercritical carbon dioxide process. *Pharmaceutical Research*, **21**: p. 524–535.
11. Huang, Y.-Y. and C.-H. Wang (2006) Pulmonary delivery of insulin by liposomal carriers. *Journal of Controlled Release*, **113**: p. 9–14.
12. Bustami, R., et al. (2000) Generation of micro particles of proteins for aerosol delivery using high pressure modified carbon dioxide. *Pharmaceutical Research*, **17**: p. 1360–1366.
13. Winters, M., et al. (1996) Precipitation of proteins in supercritical carbon dioxide. *Journal of Pharmaceutical Sciences*, **85**: p. 586–594.
14. Mitchell, J. and M. Nagel (2004) *KONA*, **22**: p. 32–65.
15. Lu, D. and A. Hickey (2005) Liposomal dry powders as aerosols for pulmonary delivery of proteins. *AAPS PharmSciTech*, **6**: p. E641–E648.

13

Macromolecular Delivery Across the Blood–Brain Barrier

Kullervo Hynynen

Summary

The delivery of macromolecules into the central nervous system (CNS) via the blood stream is seriously limited by the blood–brain barrier (BBB). Noninvasive, transient, and local image-guided blood–brain barrier disruption (BBBD) can be accomplished using focused ultrasound exposure with intravascular injection of pre-formed microbubbles. A detailed description of the method for MRI-guided focal BBBD in animals will be described in this chapter. The method may open a new era in CNS macromolecular drug delivery.

Key Words: Ultrasonics; Magnetic resonance imaging (MRI); Drug delivery systems; Ultrasound contrast agents; Blood–brain barrier.

1. Introduction

The blood–brain barrier (BBB) prevents the passage of many macromolecules from circulation into the brain parenchyma, thus seriously limiting the usefulness of these agents in the treatment of CNS diseases (*1–4*). The BBB is formed by the endothelial cells of the cerebral microvessels that connect to each other by intracellular attachments known as "tight junctions" (*1,2*). In addition, there is a physiological barrier at the level of basal lamina (*5*) that actively removes undesirable molecules from the brain. The tight junctions can be opened temporarily by an intraarterial injection of certain chemicals such as mannitol or other hyperosmotic solutions that cause the endothelial cells to shrink (*6*). Similar penetration of the BBB can be achieved by chemical modification of the macromolecules by making them lipophilic, or by using other carriers such as

From: *Methods in Molecular Biology, vol. 480: Macromolecular Drug Delivery*, Edited by: M. Belting
DOI 10.1007/978-1-59745-429-2_13, © Humana Press, a part of Springer Science+Business Media, LLC 2009

amino acids and peptides that are selectively transported through the BBB. All of these methods produce a diffused BBBD within the entire tissue volume supplied by the injected artery branch *(1,2)* without the ability to selectively target a small brain volume.

The development of noninvasive imaging methods, such as magnetic resonance imaging (MRI), allows the visualization of not only the anatomy but also the physiology, function, and location of specific molecules of the brain. This provides an opportunity to localize specific therapeutic targets in the brain, and it would be desirable to be able to deliver therapeutic agents only in these locations, thus minimizing potential harmful effects outside the target volume. This type of localized drug delivery has been accomplished by the direct injection of agents through a catheter into the targeted region of the brain *(2)*. As a less-invasive alternative, this could be accomplished by disrupting the BBB only in selected locations, leaving the intact BBB to protect the surrounding regions.

A few years ago, our group demonstrated noninvasive and reversible disruption of the BBB at targeted locations using focused ultrasound bursts in conjunction with an ultrasound contrast agent *(7)*. Later reports *(8–16)* have extended the initial findings and verified the effectiveness of the proposed method. Since ultrasound can be focused noninvasively through the skull under MRI guidance *(17)*, this method offers huge potential for research and eventually for clinical patient treatments and diagnosis. A detailed description of the method for MRI-guided focal BBBD in animals will be described in this chapter.

2. Materials

2.1. Ultrasound and MRI Contrast Agents

1. We have experience in using two different ultrasound contrast agents [Optison® (GE Healthcare, Milwaukee, WI, USA) and Definity® (Bristol-Myers Squibb Medical Imaging, N. Billerica, MA, USA)] that produce almost identical results *(18)*. Both are stored at +4°C. It is likely that similar results could be obtained with other ultrasound contrast agents containing microbubbles.
2. The disruption of the BBB with focused ultrasound can be visualized using standard commercial MRI contrast agents, such as gadopentetate dimeglumine (MAGNEVIST, Berlex Laboratories, Inc., Wayne, NJ, USA; molecular weight of 928) that do not penetrate through the BBB.

2.2. Degassed Water

1. The ultrasound beam requires a medium to propagate from the transducer to the tissue. Water is the most convenient medium for this purpose, since its acoustic attenuation is small and the speed of sound and density are close to those in soft tissues. Thus, the ultrasound beam propagates almost completely undisturbed

through water and couples well to the skin with minimal reflections. Since gas bubbles significantly disturb ultrasound propagation, it is important to minimize the potential for bubble formation. This can be done by degassing the water prior to use.

2. Degassing can be done by boiling the water for 20–30 min and then allowing it to cool to room temperature. An alternative method is to use degassing systems that remove the gas with the aid of a vacuum pump.

3. Methods

3.1. Generation of the Focused Ultrasound Exposure

3.1.1. Ultrasound Transducer

1. The ultrasound fields are generated by a focused, MRI compatible piezoelectric transducer *(19)*. The physical characteristics of the transducer determine the size of the focal spot and can be tailored for the experimental needs. For example, we have performed successful experiments with 100-mm diameter, 80-mm radius of curvature transducers with a frequency of 0.7 MHz (*see* **Note 1**).

3.1.2. RF Signal Driving the Transducer

1. The RF-signal feeding the transducer can be generated with many different devices. As an example, we have used the following commonly available laboratory system (*see* **Note 2**).
2. A frequency generator (Wavetek, Model 271, San Diego, CA, USA) is used to generate the RF signal.
3. The RF signal is amplified by an RF amplifier (Model 240L; ENI, Inc., Rochester, NY, USA).
4. The forward and reflected electric power is measured in continuous wave (CW) mode before the exposures using a digital power meter (HP Model) and a dual directional coupler (Werlatone Model C1373).

3.1.3. Ultrasound Transducer Calibration

1. The acoustic power output and the focal pressure amplitude as a function of applied RF power and driving voltage need to be measured as described earlier *(20)*.
2. Place the transducer in degassed water in a container, the walls of which are lined with sound absorbing rubber mats.
3. Position a calibrated hydrophone at focal distance from the transducer. The hydrophone is connected to a standard oscilloscope to measure its voltage. We have used a calibrated membrane hydrophone (spot diameter 0.5 mm, GEC-Marconi

Research Center, Chelmsford, England) for the whole pressure amplitude range used (*see* **Note 3**).

4. Move the hydrophone across the focal plane, while the ultrasound is pulsed, until the maximum voltage generated in the hydrophone is found.
5. Record the amplitude of the voltage (maximum and minimum) as a function of the driving RF voltage of the frequency generator.
6. Calculate the acoustic pressure amplitude values using the calibration coefficient of the hydrophone.

3.2. Exposure System

1. **Figure 1** shows a diagram of the transducer-positioning equipment used for the experiments.
2. Mount the ultrasound transducer on a mechanical arm in a plastic container filled with degassed water.
3. Place an MRI coil around the opening that allows the beam to propagate from the water to the target on top of the container.
4. Move the transducer with the positioning arm and aim the focal spot through a hole in the plastic plate, so that it will enter the head of the animal when it is lying on its back on the plate. The movement of the transducer arm can be accomplished with manual lead screw-based positioning system or a remote-controlled hydraulic or mechanical computer-controlled system such as described for MRI-guided focused ultrasound surgery *(21)* (*see* **Note 4**).

3.3. Magnetic Resonance Imaging (MRI)

3.3.1. Equipment

1. For these studies, standard MRI scanners can be used to target the brain structures and to verify the BBBD. The only requirements are that the system has adequate space for the sonication system and high enough resolution to image animal brain.

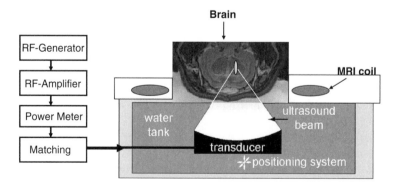

Fig. 1. A diagram of the sonication system and the equipment.

We have performed our experiments in standard clinical 1.5 and 3 T magnets (General Electric) with small surface coils that allow good images of the animal brains.

3.3.2. Pulse Sequences

1. Anatomical localization is performed by acquiring three sets of orthogonal T2-weighted fast spin echo (FSE) images (*see* **Table 1** for parameters).
2. BBBD verification and quantification is performed by acquiring a time series of T1-weighted FSE images (*see* **Table 1** for parameters).

3.3.3. Registration of the Ultrasound Beam Focus with MRI

1. Before starting the experiments, the focal spot is localized in the MRI coordinate system.
2. Aim the ultrasound beam on the water surface at a low power level (1 W).
3. Place a marker at the focal spot location that is visible as a fountain in the water.
4. Obtain an FSE T2-weigted image of the marker.
5. Measure the location of the marker in the MRI coordinate system.
6. Use this location as a reference location when targeting structures in the brain.

3.4. Animal Preparation

3.4.1. Preparation of a Skull Window

1. Anaesthetize the rabbit (approximately 3–4 kg) using a mix of 40 mg/kg ketamine (Aveco Co, Inc., Fort Dodge, IA, USA), and 10 mg/kg of xylazine (Lloyd Laboratories, Shenandoah, IA, USA).
2. Remove the hair from the skull with clippers and hair removal lotion.
3. Apply antiseptic lotion to sterilize the skin.
4. Cut the skin over the central line of the top of the skull.
5. Push the skin sideways to expose the skull surface.

Table 1
Parameters Used in the MR Imaging

Sequence	TR/TE (ms)	Flip angle (°)	Echo train length	Field of view (cm)	Matrix size	Slice thickness (mm)	Bandwidth
T2-weighted FSE	2000/85	90	8	10	256×256	1.5	16
T1-weighted FSE (contrast*)	500/17	90	4	10	256×256	1.5	16

*The T1-weighted sequence to determine the BBBD after an injection of MRI contrast agent

6. Remove a piece of skull (approximately 20 × 20 mm).
7. Replace the skin over the bone window.
8. Suture the skin back together.
9. Allow the animal to recover from the anesthesia.
10. Provide pain medication 2 days post-surgery.
11. Allow the wound to heal and then remove the sutures.
12. Allow the suture holes to heal.
13. The BBBD is performed a minimum of 10 days post-surgery.

3.4.2. Preparation of the Animal Prior to the Experiment

1. The BBBD can be induced both in animals with the skull window and intact skull provided that adequate ultrasound transmission is achieved into the brain.
2. All animals are prepared the same way for BBBD.
3. Animal is anesthetized.
4. Remove the hair over the top of the head by shaving and then by using hair removal lotion.
5. Insert an intravenous catheter in the ear (rabbit) or tail vein (mice, rats).
6. Place the body of the animal on a water blanket, through which temperature controlled water is circulated to maintain the temperature of the animal.
7. Place a rectal thermometer to monitor the temperature of the animal.

3.5. Positioning of the Animal for the Experiment

1. Place the animal on its back on the sonication system such that the top of the head is in the middle of the imaging coil over the window for the ultrasound beam.
2. Inject degassed water under the head to couple the head to the plastic membrane of the sonication system.
3. Secure the body and the head with tape or straps.
4. Move the sonication system into the magnet.
5. Perform three orthogonal sets of multi-slice T2-weighted FSE images to assure appropriate head position and good acoustic coupling.
6. Re-position the head if needed and repeat imaging.
7. Select the target locations from the images.
8. Aim the ultrasound beam focus to the first target location.

3.6. BBB Disruption

3.6.1. Microbubble Injection

1. Inject the ultrasound contrast agent containing microbubbles intravenously through the ear vein (rabbits) or tail vein (rats, mice) simultaneously with the start of each sonication. The Definity® dosage is 10 µl/kg of body weight, which is recommended by the manufacturer for clinical use. The Optison® dosage is

50 μl/kg of body weight, which is in the range (0.5–5.0 ml; i.e., 7.1–71 μl/kg for a 70-kg adult) recommended for human use (*see* **Note 5**).

2. Follow the contrast agent injection with an injection of approximately 0.5 ml/kg of saline to flush the agent from an access line (*see* **Note 6**).

3.6.2. Ultrasound Exposure

1. Start the sonication, simultaneously, with the start of the microbubble injection. The sonication has a burst length of 10 ms and a repetition frequency of 1 Hz. (*see* **Note 7**).
2. The sonication can be repeated in another location after the bubbles are cleared from the circulation system. This requires a minimum of 5 min (*see* **Note 8**).

3.7. Delivery of Macromolecular Agent

1. Most effective delivery of the macromolecular agent through the BBB is achieved when the agent is injected, simultaneously, with the microbubbles, while exposing the target volume to the ultrasound *(9)*. The macromolecular agent can be injected after the sonications and, at least for smaller molecules, some BBB penetration will remain up to approximately 6 h after the exposure.

3.8. Detection of the BBBD

1. Acquire a T1-weighted image across the brain at the focal depth (**Table 1**).
2. Inject MRI contrast agent IV (0.125 mmol/kg).
3. Repeat the T1-weighted image multiple times.
4. Subtract the signal intensity of the first image from the follow-up images.

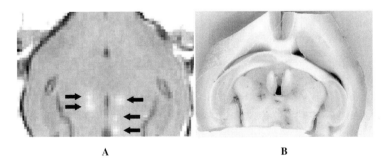

A   B

Fig. 2. (**a**) A T1-weighted contrast enhanced MR image across the beam propagation demonstrating four locations of contrast enhancement. These locations correlate with sonication locations. (**b**) A photograph of a post-mortem section of same brain at the same location demonstrating locations where Trypan Blue was leaking across the BBB. These locations correlate with the sonication locations.

5. The increase in the signal intensity at the sonicated locations, over that in the same anatomical structure on the contra-lateral side, is proportional to the size of the BBB opening and the amount of macromolecular therapyagent delivered into the brain *(14)* (*see* **Fig. 2**).

4. Notes

1. We have used several different frequencies between 0.25 and 2 MHz. The use of lower frequencies not only increases the focal spot size but also reduces the number of erythrocyte extravasations per unit area *(9)*. The reduction in the diameter of the transducer, or increase in the radius of curvature, also increases the focal spot size. Examples of the 50% focal spot dimensions for different transducers are given in **Table 2**. The electrical impedance of the transducer is, typically, matched to the output impedance of the amplifier by an external LC-matching network to allow optimum power transfer. Although, we manufactured our own transducers in-house, custom MRI compatible transducers can be purchased from several manufacturers such as Imasonic, Inc. (Besancon, France).
2. The equipment described is no longer manufactured. Other frequency generators, RF amplifiers, and power meters can be used instead. An equivalent device having all of the components integrated (the frequency generation, amplification, and power measurement) is manufactured by Advanced Surgical Systems, Inc. (Tucson, AZ, USA).
3. The hydrophone we used is no longer manufactured, but an equivalent hydrophone can be purchased, for example, from Precision Acoustics (Dorchester, UK) The reported values are often estimates for the pressure amplitude in the brain obtained by decreasing the measured water values based on ultrasound attenuation through the brain with an average amplitude attenuation coefficient of 5 Np/m/MHz *(22,23)*. If through-skull exposures are performed, then the insertion loss induced by the skull bone needs to be taken into account. There is a significant variation in the through-skull propagation from location to location and from animal to animal and this may increase as the animal gets older. Typically, the lower the

Table 2
Examples of Measured Focal Spot Dimensions (50% of the Peak Pressure Amplitude) for *F*-number 0.8 Focused Transducers

Frequency (MHz)	50% Focus width (mm)	50% Focus length (mm)
0.26	8	40
0.69	2.3	14
1.1	1.7	7.5
1.5	1	4.8
3.3	0.6	3.2

Table 3
Measured Pressure Amplitude Transmission Through Animal Skulls

Species	0.25 MHz	0.69 MHz	1.1 MHz
Rabbit	72%	34%	8.6%
Rat	85%	53%	–

The values are average values for limited measurements and indicate the pressure amplitude value that is measured when the beam propagates through the skull in water (water measurement= 100%). There are large variations from location to location and from skull to skull and these are only to be used as guidelines.

frequency the lower the losses in the skull. Therefore, it is recommended that the loss measurements be performed with skull samples from the animals used in the experiments. Examples of measured losses are given in **Table 3**.

4. The transducer is mounted on a mechanical positioning system that can move it to aim the focus at the desired location in the brain, based on the MR images. The positioner needs to be MRI compatible. This means that, all parts should be nonmagnetic. In general, plastic and other nonmetallic materials are fine. Non-magnetic metals (such as brass, copper, aluminum) can also be used, but their amount should be kept small and removed from the imaging space. Descriptions of a positioning system have been reported for experimental *(19,24)* and proto-type clinical systems *(21)* used in MRI-guided ultrasound surgery, and they are applicable for these experiments also. In addition, the computer-controlled experimental animal positioning system developed for our experiments is available for research use.

5. These ultrasound contrast agents consist of pre-formed bubbles that are either lipid (Definity®) or human serum albumin (Optison®) shells filled with the perfluoro-carbon gas (Perflutren).

6. The bubbles are cleared from the blood circulation system such that the injection can be repeated after a minimum of a 5-min interval.

7. Most of our experience is with a duration of 20 s for the whole sonication. With this short sonication duration, a 10-s delay is used with rabbits to allow the microbubbles to enter into the brain. With longer exposures, the sonication is started simultaneously with the injections. Longer sonications will not only pro-duce larger BBBD but may also produce more tissue damage. We have found that a sonication of 300 s provides close to the maximum BBBD. The peak acoustic power and pressure amplitude for the threshold of the BBBD is frequency depen-dent and is roughly proportional to $1/frequency^{-2}$. The BBBD is not enhanced by increasing the burst duration to 100 ms, but the threshold is increased when 1 ms or shorter exposures are used. The burst repetition frequency of 0.5–2 Hz provided equal results. The BBBD increases with increasing concentration of microbubbles *(14)*.

8. There are two options when multiple locations are targeted. When a manual or slow positioning device is used, then each of the locations is targeted serially with a minimum time delay of 5 min. For each sonication, the microbubble injection is repeated. When fast positioning is possible, either with a mechanical system or with phased array, then each location will be exposed to a single burst one after another during the injection, and then the sonications are repeated at a repetition frequency of 1 Hz. Thus, all of the locations will be sonicated during a single ultrasound bolus injection. This will require an automated sonication system.

Acknowledgments

Sources of support: NIH (grant numbers EB00705 and EB003268), and the Terry Fox Foundation.

References

1. Abbott, N.J., and Romero, I.A. (1996) Transporting therapeutics across the blood-brain barrier. *Mol. Med. Today* **2**,106–113.
2. Kroll, R.A., and Neuwelt, E.A. (1998) Outwitting the blood-brain barrier for therapeutic purposes: osmotic opening and other means. *Neurosurgery* **42**,1083–1099.
3. Pardridge, W.M. (2002) Drug and gene delivery to the brain: the vascular route. *Neuron* **36**,555–558.
4. Nag, S. (2003) Morphology and molecular properties of cellular components of normal cerebral vessels. *Methods Mol. Med.* **89**,3–36.
5. Muldoon, L.L., Pagel, M.A., Kroll, R.A., Roman-Goldstein, S., Jones, R.S., and Neuwelt, E.A. (1999) A physiological barrier distal to the anatomic blood-brain barrier in a model of transvascular delivery [see comments]. *AJNR Am. J. Neuroradiol.* **20**,217–222.
6. Doolittle, N.D., Miner, M.E., Hall, W.A., Siegal, T., Jerome, E., Osztie, E., McAllister, L.D., Bubalo, J.S., Kraemer, D.F., Fortin, D., Nixon, R., Muldoon, L.L., and Neuwelt, E.A. (2000) Safety and efficacy of a multicenter study using intraarterial chemotherapy in conjunction with osmotic opening of the blood-brain barrier for the treatment of patients with malignant brain tumors. *Cancer* **88**, 637–647.
7. Hynynen, K., McDannold, N., Vykhodtseva, N., and Jolesz, F.A. (2001) Noninvasive MR imaging-guided focal opening of the blood-brain barrier in rabbits. *Radiology* **220**,640–646.
8. Hynynen, K., McDannold, N., Sheikov, N.A., Jolesz, F.A., and Vykhodtseva, N. (2005) Local and reversible blood-brain barrier disruption by noninvasive focused ultrasound at frequencies suitable for trans-skull sonications. *Neuroimage* **24**, 12–20.
9. Hynynen, K., McDannold, N., Vykhodtseva, N., Raymond, S., Weissleder, R., Jolesz, F.A., and Sheikov, N. (2006) Focal disruption of the blood-brain barrier due to 260-kHz ultrasound bursts: a method for molecular imaging and targeted drug delivery. *J. Neurosurg.* **105**,445–454.
10. Sheikov, N., McDannold, N., Vykhodtseva, N., Jolesz, F., and Hynynen, K. (2004) Cellular mechanisms of the blood-brain barrier opening induced by ultrasound in presence of microbubbles. *Ultrasound Med. Biol.* **30**,979–989.

11. Sheikov, N., McDannold, N., Jolesz, F., Zhang, Y.Z., Tam, K., and Hynynen, K. (2006) Brain arterioles show more active vesicular transport of blood-borne tracer molecules than capillaries and venules after focused ultrasound-evoked opening of the blood-brain barrier. *Ultrasound Med. Biol.* **32**,1399–1409.

12. Kinoshita, M., McDannold, N., Jolesz, F.A., and Hynynen, K. (2006) Targeted delivery of antibodies through the blood-brain barrier by MRI-guided focused ultrasound. *Biochem. Biophys. Res. Commun.* **340**,1085–1090.

13. Kinoshita, M., McDannold, N., Jolesz, F.A., and Hynynen, K. (2006) Noninvasive localized delivery of Herceptin to the mouse brain by MRI-guided focused ultrasound-induced blood-brain barrier disruption. *Proc. Natl. Acad. Sci. U.S.A.* **103**,11719–11723.

14. Treat, L.H., McDannold, N., Vykhodtseva, N., Zhang, Y., Tam, K., and Hynynen, K. (2007) Targeted delivery of doxorubicin to the rat brain at therapeutic levels using MRI-guided focused ultrasound. *Int. J. Cancer* **121**,901–907.

15. Choi, J.J., Pernot, M., Small, S.A., and Konofagou, E.E. (2007) Noninvasive, transcranial and localized opening of the blood-brain barrier using focused ultrasound in mice. *Ultrasound Med. Biol.* **33**,95–104.

16. Yang, F.Y., Fu, W.M., Yang, R.S., Liou, H.C., Kang, K.H., and Lin, W.L. (2007) Quantitative evaluation of the use of microbubbles with transcranial focused ultrasound on blood-brain-barrier disruption. *Ultrasound Med. Biol.* **33**,1421–1427.

17. Hynynen, K., Clement, G.T., McDannold, N., Vykhodtseva, N., King, R., White, P.J., Vitek, S., and Jolesz, F.A. (2004) 500-element ultrasound phased array system for noninvasive focal surgery of the brain: A preliminary rabbit study with ex vivo human skulls. *Magn. Reson. Med.* **52**,100–107.

18. McDannold, N., Vykhodtseva, N., and Hynynen, K. (2007) Use of ultrasound pulses combined with definity for targeted blood-brain barrier disruption: a feasibility study. *Ultrasound Med. Biol.* **33**,584–590.

19. Hynynen, K., Darkazanli, A., Unger, E., and Schenck, J.F. (1993) MRI-guided noninvasive ultrasound surgery. *Med. Phys.* **20**,107–115.

20. Hynynen, K., Vykhodtseva, N.I., Chung, A., Sorrentino, V., Colucci, V., and Jolesz, F.A. (1997) Thermal effects of focused ultrasound on the brain: determination with MR Imaging. *Radiology* **204**,247–253.

21. Cline, H.E., Hynynen, K., Watkins, R.D., Adams, W.J., Schenck, J.F., Ettinger, R.H., Freund, W.R., Vetro, J.P., and Jolesz, F.A. (1995) A focused ultrasound system for MRI guided ablation. *Radiology* **194**,731–737.

22. Goss, S.A., Johnson, R.L., and Dunn, F. (1978) Comprehensive compilation of empirical ultrasonic properties of mammalian tissues. *J. Acoust. Soc. Am.* **64**, 423–457.

23. Goss, S.A., Johnson, R.L., and Dunn, F. (1980) Compilation of empirical ultrasonic properties of mammalian tissues. II. *J. Acoust. Soc. Am.* **68**,93–108.

24. Cline, H.E., Schenck, J.F., Watkins, R.D., Hynynen, K., and Jolesz, F.A. (1993) Magnetic resonance guided thermal surgery. *Magn. Reson. Med.* **31**,628–636.

14

Positron Emission Tomography (PET) and Macromolecular Delivery In Vivo

Ludwig G. Strauss and Antonia Dimitrakopoulou-Strauss

Summary

Positron emission tomography (PET) examinations with F-18-fluorodeoxyglucose (FDG) provide detailed information about the glucose-like metabolism in tissue. It is generally accepted that FDG reflects the viability of tumour cells. The kinetics of FDG is modulated by several genes, besides the glucose transporters and hexokinases. Additional specific information can be obtained non-invasively by using other tracers specific for cell membrane receptors. PET studies with radiolabelled peptides have emerged as a new diagnostic tool for imaging of certain tumour entities, like neuroendocrine tumours (NETs) and gastrointestinal stromal tumours (GISTs). This application is based on certain properties of these tumours, like the overexpression of somatostatin receptors, which can be visualised by somatostatin analogues, like 1,4,7,10-tetraazacyclododecane-N, N', N'', N'''-tetraacetic-acid-D-Phe1-Tyr3 octreotide (DOTATOC) in NET. The overexpression of gastrin-releasing peptide (GRP) receptors can be visualised in GIST by using bombesin analogues. These peptides can be labelled by ^{68}Ga, which is a generator product and therefore more cost-effective than cyclotron products. ^{68}Ga-DOTATOC is a peptide that binds primarily to somatostatin receptor subtype 2 (SSTR2). PET studies with ^{68}Ga-DOTATOC are performed in patients with NET and some other tumours. ^{68}Ga-BZH3 (^{68}Ga-Bombesin) is a peptide that binds to at least three bombesin receptor subtypes: the BB1 (also known as neuromedin B), the BB2 (also known as GRP), and the BB3 (bombesin receptor subtype 3). This bombesin analogue, ^{68}Ga-BZH3, is used in patients with GIST.

Key Words: FDG; ^{68}Ga-DOTATOC; ^{68}Ga-Bombesin; NET; GIST; PET; SSTR2.

From: *Methods in Molecular Biology, vol. 480: Macromolecular Drug Delivery*, Edited by: M. Belting
DOI 10.1007/978-1-59745-429-2_14, © Humana Press, a part of Springer Science+Business Media, LLC 2009

1. Introduction

Positron emission tomography (PET) is a promising method for the evaluation of pathophysiological processes like cellular metabolism, tumour perfusion,
and expression of receptors in tumours. The most commonly used radiotracer for
PET examinations is ^{18}F-fluorodeoxyglucose (FDG), a glucose analogue which
reflects the tumour viability based on the increased glycolysis in malignancies.
However, the FDG uptake may only be moderately increased in slowly growing
tumours with a low proliferation rate like soft tissue sarcomas and neuroendocrine tumours (NETs) *(1–2)*. One major advantage of the nuclear medicine
procedures is the capability to enhance the specificity of diagnostics by the use
of dedicated radiopharmaceuticals. Recently, especially, receptor binding tracers have found major attention in the assessment of an enhanced expression of
receptors in tumours, thus providing non-invasively specific molecular biological information.

Radiolabelled peptides are of increasing interest in imaging of tumours.
A new somatostatin analogue, the 1,4,7,10-tetraazacyclododecane-*N*, *N'*, *N''*,
N'''-tetraacetic-acid-D-Phe1-Tyr3 octreotide (DOTATOC), labelled with ^{68}Ga is
recently available. ^{68}Ga is produced from a generator system and therefore convenient to use. ^{68}Ga-DOTATOC binds primarily to the somatostatin receptor
subtype two (SSTR2) in NETs. In vitro binding and in vivo experiments in
rats with [DOTA0,Tyr3]octreotide radiolabelled either with ^{111}In or ^{90}Y showed
favourable binding and biodistribution characteristics with high uptake and
retention in receptor-positive organs and tumour in rats. Moreover, receptor-
specific and time-dependent internalisation of [DOTA0,Tyr3]octreotide has been
demonstrated in several SSTR2 positive cell lines. The short half-life of ^{68}Ga
(68.3 min) and the fast accumulation of small peptides render ^{68}Ga-DOTATOC
to be an ideal radiotracer for receptor imaging *(3–5)*. Besides the diagnostic use,
DOTATOC can be labelled with Y-90, a β-emitter, and then be used for the treatment of endocrine tumours. Due to the specific information about an enhanced
expression of SSTR2, ^{68}Ga-DOTATOC is used for differential diagnostic purpose in several tumour types, which are known to have an enhanced receptor
expression. In patients, this tracer provides non-invasive information about gene
expression levels. Furthermore, ^{68}Ga-DOTATOC is also used to select those
patients with an enhanced receptor density for treatment with ^{90}Y-DOTATOC,
because only in patients with a high-receptor density radioisotope treatment can
be performed *(3)*.

Another peptide, which gains increasing interest is the pan-Bombesin analogue BZH3, which can be labelled with the positron emitter ^{68}Ga *(6)*. BZH3
binds to at least three different receptor subtypes: the BB1 (also known as neuromedin B), the BB2 (also known as gastrin-releasing peptide or GRP), and the

BB3 (bombesin receptor subtype 3). Reubi et al. reported in a recent paper that gastrointestinal stromal tumours (GISTs) expressed bombesin subtype 2 receptors, better known as GRP receptors, using in vitro receptor autoradiography (7). GISTs are tumours that usually occur in the wall of the bowel and are thought to be derived from the cells of Cajal that drive peristalsis in the intestine. The authors report that not only primary GISTs, but also metastases and even tumour samples of patients who did not respond to Glivec treatment were receptor positive. Based on these data, ^{68}Ga-Bombesin, in terms of a pan-Bombesin analogue, the peptide BZH$_3$, can be used in patients with GIST for differential diagnostic purpose additional to the standard tracer FDG.

A dramatic improvement in survival was noted after the discovery of Imatinib mesylate that led to a response of 50% in patients with non-resectable GIST and to stable disease in 28% (8–9). However, it is known that not all patients respond to Imatinib mesylate and that a considerable number of patients who initially responded to the treatment may become resistant later on. PET studies with both FDG and ^{68}Ga-Bombesin may be helpful to identify those patients with resistance early in the therapy follow-up period.

PET tracer studies are frequently performed as static acquisitions following a dedicated time after tracer injection. In this way, the overall tracer uptake can be quantified in the regions of interest. However, the tracer uptake is the sum of very different effects, including the vessel density of the tumour, transport of a metabolically active tracer into the cells, and the further intracellular metabolism of a radiopharmaceutical. In receptor active tracers, besides the fractional blood volume the receptor binding and release as well as the internalisation of the tracer is important for the global uptake noted in the PET images. Therefore, a dynamic data acquisition focussed on the primary volume of interest, followed by additional static acquisitions after repositioning of the patient provides the possibility to apply compartment and non-compartment models to the dynamic data and to obtain more detailed information about the biological processes. Furthermore, due to the additional static acquisitions whole body images can be reconstructed. However, sophisticated software packages are needed to facilitate the data analysis.

2. Materials

2.1. Radiopharmaceuticals

1. FDG is either obtained from an external production site or produced at the PET site, if a cyclotron is available. The production itself is usually done with one of the commercially available automated synthesisers according to the outlines of the manufacturer.

2. ^{68}Ga (half-life, 68.3 min; β+ (beta plus), 88%; Eβ+ maximum, 1.900 keV) is obtained in 0.5 mL of 0.5 N HCl from a ^{68}Ge/^{68}Ga radionuclide generator developed by the Radiochemistry Department of the German Cancer Research Center.
3. DOTAO-DPhe1-Tyr3 octreotide is synthesised at our centre according to the method described by Heppeler et al. *(10)*.

2.2. Columns

1. Small-sized anion exchanger column (e.g. AG 1×8 Cl⁻, mesh 200–400; Bio-Rad).
2. Silica gel-based reversed phase cartridge (SepPAK; Waters Corp., Milford, MA, USA).

3. Methods
3.1. Tracer Production

1. ^{68}Ga for PET is obtained from a ^{68}Ge/^{68}Ga generator, which usually consists of a column containing a phenolic ion-exchanger loaded with ^{68}Ge and coupled in series with a small-sized anion exchanger column to concentrate ^{68}Ga during elution *(11)*. A generator may provide ^{68}Ga with an average yield of approximately 60% for >1.5 years.
2. DOTAO-D-Phe1-Tyr3 octreotide is synthesised at our centre according to the method described by Heppeler et al. *(10)*. Following evaporation of the HCl, ^{68}Ga is redissolved in 200 μl of 0.1 *M* acetate buffer pH 4.8 and 15 μl of 1 m*M* aqueous DOTATOC solution is added. The mixture is kept for 15 min at 90°C. Non-complexed ^{68}Ga is separated from ^{68}Ga-DOTATOC using a silica gel-based reversed phase cartridge equilibrated with 0.1 *M* acetate buffer pH 6.5. While uncomplexed ^{68}Ga is retained in the column, ^{68}Ga-DOTATOC can be eluted with 1.5 ml ethanol. After evaporation of the organic solvent, the compound is redissolved in 5 ml of 0.01 *M* phosphate buffered saline (PBS). Pyrogenicity and sterility are also checked. Paper chromatography usually indicates less than 2% of unchelated ^{68}Ga in the final preparation.
3. Bombesin (BZH$_3$) labelling is performed with Ga-68 using the procedure described by Schuhmacher et al. *(6)*. For peptide labelling, the generator eluate containing 0.5 GBq of ^{68}Ga in 0.2 ml of 0.5 *M* HCl is evaporated to dryness and redissolved in 0.2 ml of 0.1 *M* acetate buffer (pH 4.8). After the addition of 5 L of 1 m*M* aqueous solution of BZH$_3$, the mixture is kept for 10 min at 90°C. Uncomplexed ^{68}Ga is separated by adsorption onto a C18-coated silica gel cartridge that is equilibrated with 0.1 *M* acetate buffer (pH 6.2), whereas ^{68}Ga-BZH$_3$ can be eluted with 1.5 ml ethanol. After evaporation of the organic solvent, the compound is redissolved in 0.01 *M* PBS pH 7, containing 0.5 mg/ml human serum albumin. The preparations are checked for bound and free ^{68}Ga by paper chromatography using Whatman no. 1 and a mixture of methanol and 0.01 *M* acetate buffer (pH 6.2) in a ratio of 55:45. ^{68}Ge contamination of the ^{68}Ga-BZH$_3$ preparations

is determined by counting after a waiting period of 30 h, which ensures complete ^{68}Ga decay.

3.2. PET Studies

3.2.1. PET Systems

1. Dedicated PET systems are usually based on full ring detector systems with an axial field of view exceeding 15 cm, and can be operated in septa extended (2D mode) or septa retracted mode (3D) for patient examinations. Some systems only provide 3D-data acquisition modes. Most systems allow the simultaneous acquisition of 36 transversal slices and more with a theoretical slice thickness of 2–5 mm. Transmission scans for a total of 10 min are obtained prior to the radionuclide application, for the attenuation correction of the acquired emission tomographic images. All PET images are attenuation corrected and iteratively reconstructed.
2. PET/CT systems are gaining major attention in the last few years. Due to the sequential acquisition of CT and PET data, morphological and functional information can be easily correlated and analysed. The CT data can also be used for attenuation correction of the PET images, thus decreasing the total examination time of the patient.
3. PET/MRI systems are coming up at the horizon of the recent developments. Currently, an experimental PET/MRI system is evaluated at a few sites, for the use of brain studies in patients.

3.2.2. Acquisition

1. It is important for ^{18}F-FDG studies that all patients have been fastened for at least 4 h before PET (*see* **Notes 1 and 6**). Furthermore, blood glucose levels should be checked prior to PET to assure that they are within normal range ($<$130 mg/dl). Dynamic PET studies are performed for 60 min following the intravenous application of 300–370 MBq ^{18}F-FDG in adult patients. The frame rates are different at different PET centres. We usually perform a 28-frame protocol (10 frames of 30 s, 5 frames of 60 s, 5 frames of 120 s, and 8 frames of 300 s), which covers a time range of 1 h. Dynamic studies are preferred in patients examined with PET for therapy management. Static studies include transmission and emission scanning for at least 3 (transmission) and 5 (emission) min for each bed position. The procedure is different for PET/CT systems, because the CT is usually acquired for the total scanning range separately from the PET data acquisitions.
2. ^{68}Ga-DOTATOC (150–210 MBq) or ^{68}Ga-Bombesin (3 nmol) is usually injected into adult patients for receptor imaging (*see* **Notes 2–4**). The acquisition protocols are the same as in FDG. If different tracer studies are planned, the studies should be performed at different days to avoid background activity due to the decay of the injected isotope.

3.2.3. Evaluation of PET Dynamic Data

The basic evaluation includes the visual assessment of whole body images, using a 3D viewer, and the analysis of maximum intensity projection (MIP) images. Furthermore, a quantitative analysis of suspicious lesions and reference areas is recommended.

3.2.3.1. STANDARDISED UPTAKE VALUES (SUVs)

The reconstructed images reflect counts per volume or radioactivity concentrations. The data are converted to SUV images based on the following formula *(12)*:

$$SUV = [\text{tissue concentration (Bq/g)}]/[\text{injected dose (Bq)/body weight (g)}]$$

SUV provides an easy approach to obtain distribution values in order to compare the results from different PET examinations or patient studies. SUV has no dimension and is related to the increased or decreased accumulation of the tracer in a local region, as compared to the average radioactivity concentration based on the individual dose and body weight. Usually SUV measured 55–60 min post-injection is used for the analysis of tracer studies.

3.2.3.2. DYNAMIC DATA EVALUATION

The evaluation of the dynamic PET data demands the use of appropriate software to apply compartment and non-compartment models to the dynamic data (**Fig. 1**). First, time–activity curves must be obtained for the target areas by placing regions-of-interest (ROI) in several slices, thus obtaining a volume-of-interest (VOI) for each target region. VOIs are superior to ROIs due to the better statistics. Furthermore, irregular VOIs should be used and the use of circular or elliptic ROIs should be avoided in order to cover the target area as accurately as possible.

3.2.3.3. COMPARTMENT ANALYSIS

A detailed quantitative evaluation of tracer kinetics requires the use of compartment modelling. A two-tissue compartment model is used for the evaluation of the dynamic studies. This is a standard methodology in particular for the quantification of dynamic FDG studies *(13–15)*. Concerning the ^{68}Ga-peptide kinetics, again the two-tissue compartment model is applied to the data. In case of ^{68}Ga-labelled peptides, DOTATOC or bombesin, k1 is associated with the receptor binding, k2 with the displacement from the receptor, k3 with the cellular internalisation, and k4 with the externalisation. The fractional blood volume

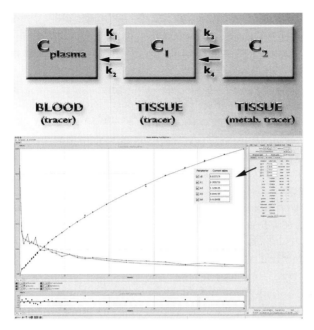

Fig. 1. Time–activity data obtained by VOI for an arterial vessel (descending curve) and a malignant tumour (ascending curve) following FDG injection. A two-compartment model is fitted to the tumour data and the results are displayed; v_B: fractional blood volume, vessel density: k1, k2: transport constants, k3, k4: constants for the phosporylation and dephosphorylation of the intracellular FDG.

(v_B), which is associated with the volume of blood exchanging with tissue, is calculated as a separate parameter *(16)*.

Compartment models usually demand an input function to obtain the arterial concentration of the tracer. While this was achieved by arterial blood sampling in previous years, several studies have shown that the input function can be derived from the images of the dynamic series. For the input function, the mean value of the VOI data obtained from a large arterial vessel like the descending aorta is used. The aorta is superior to the right ventricle, because it reflects more accurately the arterial concentration of the tracer. A vessel VOI consists of at least 7 ROIs in sequential PET images. The descending aorta is preferentially used for this purpose, because the spillover from other organs is low and the descending aorta extends from the upper chest to the lower abdomen. The recovery coefficient should be determined once for each PET system. The coefficient is dependent on both the PET hardware as well as the reconstruction parameters. The recovery coefficient is 0.85 for a diameter of 8 mm and for iteratively reconstructed images of our PET system. Noise in the input curve has an effect on the parameter estimates. Therefore, a preprocessing tool, which allows a fit of the input curve, namely by a sum of up to three decaying exponentials to reduce noise can be helpful. We are using the PMod software (PMOD Technologies Ltd., Adliswil, Switzerland) for data evaluation.

One major advantage of the PMod software is the graphical interface that allows the interactive configuration of the kinetic model by the user as well as the application of some preprocessing steps, e.g. setting up initial values and limits for the fit parameters. Visual evaluation of each plot is performed to check the quality of each fit. Each model curve is compared with the corresponding time–activity curve and the total X^2 difference was used as the cost function, where the criterion was to minimise the summed squares (X^2) of the differences between the measured and the model curve. The distribution at each individual point is taken to be Gaussian, with a standard deviation to be specified. The model parameters are usually accepted when k1–k4 is less than one and the v_B value exceeds zero. The unit for the rate constants k1–k4 is 1/min, while v_B reflects the fraction of blood within the evaluated volume.

Based on the two-compartment model data, the global influx of a tracer can be calculated by influx = k1 × k3/(k2 + k3). Another method to calculate the influx is the use of a graphical method, which requires only an input and target time–activity curve.

3.2.3.4. NON-COMPARTMENT ANALYSIS

Besides the compartment analysis, non-compartment models can be used. One frequently used procedure is the regression method. This method performs a linear regression fit on a voxel basis. The slope image provides information about the trapping of the tracer, while the intercept image reflects the distribution volume of the radiopharmaceutical. Another non-compartment model is based on the calculation of the fractal dimension (FD) *(17)*. FD is a parameter for the heterogeneity and is calculated for the time–activity data of each individual VOI. The values of FD vary from 0 to 2 showing the deterministic or chaotic distribution of the tracer activity. We use a subdivision of 7 × 7 and a maximal SUV of 20 for the calculation of FD.

Generally, all analytical methods can be applied for VOIs or on the voxel level to calculate the parametric images. These images reflect certain properties of the tracer within their spatial distribution. The parametric images can be helpful to support diagnostic or therapeutic decisions. **Figure 2** demonstrates a comparison of the different parametric imaging methods.

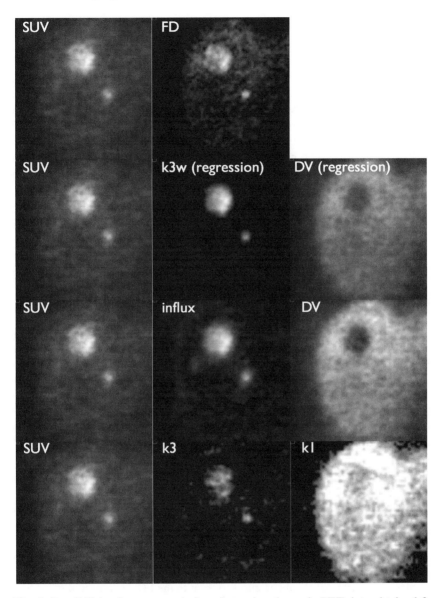

Fig. 2. Possibilities for parametric imaging using dynamic PET data obtained from an FDG examination. The cross-section of the liver demonstrates two metastases from a patient with a breast carcinoma. The parametric images demonstrate that in this case the global FDG uptake is mainly related to an enhanced intracellular phosphorylation of the tracer and less dependent on the transport of FDG into the cells; *upper row*: SUV image (SUV), fractal dimension (FD); *second row*: SUV image (SUV), k3-weighted image (k3w), and distribution volume (DV), calculated using a non-compartment

4. Notes

1. The most commonly used radiopharmaceutical is FDG, which is intravenously injected into patients fasting for at least 4 h prior to PET. Data are acquired either dynamically following tracer injection or as static images 1 h after FDG application. Whole body images can be generated if the patient is repositioned, and overlapping volumes are examined. FDG is used primarily not only in malignant tumours, but also for some non-malignant diseases like osteomyelitis *(12)*. Furthermore, FDG is an established tracer in cardiology and neurology.

2. ^{68}Ga-DOTATOC is primarily used in endocrine tumours, for both tumour diagnostics and therapy management by assessing the receptor density prior to treatment with ^{90}Y-DOTATOC. Furthermore, ^{68}Ga-DOTATOC has found to be useful for differential diagnostic reasons in some tumour types like meningiomas and lung tumours. An enhanced expression of SSTR2 had also been reported for medullary thyroid carcinoma and some lymphomas.

3. Nearly, all NETs show an enhanced binding of ^{68}Ga-DOTATOC prior to any treatment. With the progress of the disease and a longer chemotherapeutic treatment period, the SSTR2 expression can decrease and sometimes lesions may be recognised with a low accumulation of ^{68}Ga-DOTATOC or even negative findings. Following ^{90}Y-DOTATOC treatment, the global tracer uptake exhibits frequently no major changes. However, recent compartment data provided evidence of a primary effect on the vessel density, which is decreasing following radioisotope treatment.

4. ^{68}Ga-Bombesin had been used in a few tumour types, including prostate cancer, breast cancer, small cell lung tumours, and colorectal cancer. However, one promising application is the use in GISTs. PET examinations in GIST are performed prior to treatment for diagnostic purpose or following the treatment with Glivec. Additionally, a PET-FDG study may be performed to assess the tumour viability.

5. Quality control protocols should be performed for the PET system on a regular basis. Furthermore, it is helpful to acquire a blank scan prior to the first patient examination each day. For this purpose, a transmission scan is initiated to assess differences in detector sensitivity. We use this scan for the correction of the patient data on a daily basis.

6. The quantitative assessment of PET data demands the accurate measurement of the injected dose (e.g. considering residual activity in the syringe) and the measurement of the body weight of each patient. Furthermore, for FDG scans the blood glucose level should be checked prior to the FDG application.

Fig. 2. (Continued) method (regression model); *third row*: SUV image (SUV), global metabolic rate (influx), and distribution volume (DV). The parametric images are obtained by applying the Patlak model to the data; *fourth row*: SUV image (SUV), phosphorylation rate (k3), and transport rate (k1). The images are obtained by a voxel-based application of the two-compartment model.

7. The two-compartment model usually requires iterative curve fits. The critical part of iterative fits is the selection of initial values for v_B and k1–k4. Usually values in the range of 0.05–0.15 work well. Restrictions should be used to avoid values exceeding one for k1–k4. In most cases, it is helpful to perform the compartment fitting in several steps, e.g. by selecting a fit for v_B and k1 first, followed by k3, etc., in order to obtain stable, reproducible results.

Acknowledgements

We like to thank Mr. Christian Schoppa for his help with the PET studies and Mr. Armin Runz and Mr. Martin Schäfer for their help with the peptide synthesis.

References

1. Dimitrakopoulou-Strauss A, Strauss LG, Schwarzbach M, et al. (2001) Dynamic PET ^{18}F-FDG studies in patients with primary and recurrent soft-tissue sarcomas: impact on diagnosis and correlation with grading. *J Nucl Med* **42,** 713–720.
2. Adams S, Baum R, Rink T, Schumm-Dräger PM, Usadel KH, Hör G (1989) Limited value of fluorine-18 fluorodeoxyglucose positron emission tomography for the imaging of neuroendocrine tumors. *Eur J Nucl Med Mol Imaging* **25,** 79–83.
3. Koukouraki S, Strauss LG, Georgoulias V, et al. (2006) Comparison of pharmacokinetics of ^{68}Ga-DOTATOC and ^{18}F-FDG in patients with metastatic neuroendocrine tumors scheduled for ^{90}Y-DOTATOC therapy. *Eur J Nucl Med Mol Imaging* **33,** 1115–1122.
4. Henze M, Dimitrakopoulou-Strauss A, Milker-Zabel S, et al. (2005) Characterization of ^{68}Ga-DOTA-D-PHE1-TYR3-Octreotide kinetics in patients with meningiomas. *J Nucl Med* **46,** 763–769.
5. Dimitrakopoulou-Strauss A, Georgoulias V, Eisenhut M, et al. (2006) Quantitative assessment of SSTR2 expression in patients with non-small cell lung cancer using ^{68}Ga-DOTATOC PET and comparison to ^{18}F-FDG PET. *Eur J Nucl Med Mol Imaging* **33,** 823–830.
6. Schuhmacher J, Zhang H, Doll J, et al. (2005) GRP receptor-targeted PET of a rat pancreas carcinoma xenograft in nude mice with a (68) Ga-labeled Bombesin (6-14) analog. *J Nucl Med* **46,** 691–699.
7. Reubi JC, Körner M, Waser B, Mazzucchelli L, Guillou L (2004) High expression of peptide receptors as a novel target in gastrointestinal stromal tumors. *Eur J Nucl Med Mol Imaging* **31,** 803–810.
8. Demetri GD, von Mehren M, Blanke CD, et al. (2002) Efficacy and safety of imatinib mesylate in advanced gastrointestinal stromal tumors. *N Engl J Med* **347,** 472–480.
9. Joensuu H, Roberts PJ, Sarlomo-Rikala M, et al. (2001) Effect of the tyrosine kinase inhibitor STI571 in a patient with a metastatic gastrointestinal stromal tumor. *N Engl J Med* **344,** 1052–1056.
10. Heppeler A, Froidevaux S, Mäcke HR, et al. (1999) Radiometal-labeled macrocyclic chelator-derivatised somatostatin analogue with superb tumor-targeting

properties and potential for receptor-mediated internal radiotherapy. *Chem Eur J* **5,** 1974–1981.

11. Schuhmacher J, Maier-Borst W (1981) A new ^{68}Ge/^{68}Ga radiosotope generator system for production of ^{68}Ga in dilute HCl. *Appl Radiat Isot* **32,** 31–36.

12. Strauss LG, Conti PS (1991) The applications of PET in clinical oncology. *J Nucl Med* **32,** 623–648.

13. Miyazawa H, Osmont A, Petit-Taboue MC, et al. (1993) Determination of ^{18}F-fluoro-2-deoxy-D-glucose rate constants in the anesthetized baboon brain with dynamic positron tomography. *J Neurosci Methods* **50,** 263–272.

14. Sokoloff L, Smith CB (1983) Basic principles underlying radioisotopic methods for assay of biochemical processes in vivo. In: Greitz T, Ingvar DH, Widén L, eds. The Metabolism of the Human Brain Studied with Positron Emission Tomography. New York, USA: Raven Press; pp. 123–148.

15. Burger C, Buck A (1997) Requirements and implementations of a flexible kinetic modeling tool. *J Nucl Med* **38,** 1818–1823.

16. Dimitrakopoulou-Strauss A, Hohenberger P, Haberkorn U, Mäcke HR, Eisenhut M, Strauss LG (2007) ^{68}Ga-labeled bombesin studies in patients with gastrointestinal stromal tumors: comparison with ^{18}F-FDG. *J Nucl Med* **48,** 1245–1250.

17. Dimitrakopoulou-Strauss A, Strauss LG, Mikolajczyk K, et al. (2003) On the fractal nature of dynamic positron emission tomography (PET) studies. *World J Nucl Med* **2,** 306–313.

Index

Printed in the United States of America